# Modern and Interdisciplinary Problems in Network Science

## A Translational Research Perspective

T0173908

# Modern and Interdisciplinary Problems in Network Science

## A Translational Research Perspective

Edited by

### Zengqiang Chen

College of Computer and Control Engineering,
Nankai University, Tianjin, P. R. China

### Matthias Dehmer

Production and Operations Management,
University of Applied Sciences,
Upper Austria, Steyr, Austria

UMIT, Department of Mechatronics and
Biomedical Computer Science, Tyrol, Austria

College of Computer and Control Engineering,
Nankai University, Tianjin, P. R. China

### Frank Emmert-Streib

Department of Signal Processing,
Tampere University of Technology, Finland

### Yongtang Shi

Center for Combinatorics and LPMC,
Nankai University, Tianjin, P. R. China

**CRC Press**
Taylor & Francis Group
Boca Raton London New York

CRC Press is an imprint of the
Taylor & Francis Group, an **informa** business

CRC Press
Taylor & Francis Group
6000 Broken Sound Parkway NW, Suite 300
Boca Raton, FL 33487-2742

First issued in paperback 2020

© 2019 by Taylor & Francis Group, LLC
CRC Press is an imprint of Taylor & Francis Group, an Informa business

No claim to original U.S. Government works

ISBN 13: 978-0-367-65706-2 (pbk)
ISBN 13: 978-0-8153-7658-3 (hbk)

**Visit the Taylor & Francis Web site at**
**http://www.taylorandfrancis.com**

**and the CRC Press Web site at**
**http://www.crcpress.com**

# Contents

# About the Editors

**Zengqiang Chen** works at the Department of Automation at Nankai University, where he is currently a professor. His research interests are in complex networks, multi-agent systems, computer application systems, nonlinear dynamic control, intelligent computing, and stochastic analysis.

**Matthias Dehmer** is a professor at University of Applied Sciences Upper Austria, Campus Steyr and University of Applied Sciences Upper Austria and UMIT—The Health and Life Sciences University. He also holds a guest professorship at Nankai University. His research interests are in graph theory, complex networks, complexity, machine big data, analytics, and information theory. In particular, he is also working on machine learning-based methods to design new data analysis methods for solving problems in manufacturing and production.

**Frank Emmert-Streib** is a professor at Tampere University Technology, Finland, in the Department of Signal Processing. His research interests are in the field of computational biology, data science, and analytics in the development and application of methods from statistics and machine learning for the analysis of big data from genomics, finance, and business.

**Yongtang Shi** is a professor at the Center for Combinatorics of Nankai University. His research interests are in graph theory and its applications, especially the applications of graph theory in mathematical chemistry, computer science, and information theory.

# *List of Editors*

**Zengqiang Chen**
College of Computer and Control Engineering, Nankai University, Tianjin, P. R. China.

**Matthias Dehmer**
Production and Operations Management, University of Applied Sciences, Upper Austria, Steyr, Austria.

and

UMIT, Department of Mechatronics and Biomedical Computer Science, Tyrol, Austria.

and

College of Computer and Control Engineering, Nankai University, Tianjin, P. R. China.

**Frank Emmert-Streib**
Department of Signal Processing, Tampere University of Technology, Finland.

**Yongtang Shi**
Center for Combinatorics and LPMC, Nankai University, Tianjin, P. R. China.

# Contributors

**Mohammed Ali Al-garadi**
Faculty of Computer Science and
     Information Technology
University of Malaya
Kuala Lumpur, Malaysia

**Graham Bent**
IBM Research
Hursley Park
Hants, United Kingdom

**Petre Caraiani**
Institute for Economic Forecasting
Romanian Academy
Bucharest, Romania

**Zengqiang Chen**
College of Computer and Control
     Engineering
Nankai University
Tianjin, P. R. China

**Matthias Dehmer**
Production and Operations
     Management
University of Applied Sciences
Upper Austria, Steyr, Austria
and
UMIT
Department of Mechatronics and
     Biomedical Computer Science
Tyrol, Austria

and

College of Computer and Control
     Engineering
Nankai University
Tianjin, P. R. China

**Martin Hromada**
Faculty of Applied Informatics
Tomas Bata University in Zlín
Zlín, Czech Republic

**Jaan Kalda**
Department of Cybernetics
Tallinn University of Technology
Tallinn, Estonia

**Valia Mitsou**
IRIF CNRS
Université Paris Diderot
Paris, France

**Abbe Mowshowitz**
Department of Computer Science
The City College of New York
New York City, New York

**Ghulam Mujtaba**
Faculty of Computer Science and
     Information Technology
University of Malaya
Kuala Lumpur, Malaysia

**Henry Friday Nweke**
Faculty of Computer Science and
    Information Technology
University of Malaya
Kuala Lumpur, Malaysia

**David Rehak**
Faculty of Safety Engineering
VSB—Technical University of
    Ostrava
Ostrava, Czech Republic

**Pavel Senovsky**
Faculty of Safety Engineering
VSB—Technical University of
    Ostrava
Ostrava, Czech Republic

**Yongtang Shi**
Center for Combinatorics and LPMC
Nankai University
Tianjin, P. R. China

**Stephanie Rendón de la Torre**
Department of Cybernetics
Tallinn University of Technology
Tallinn, Estonia

**Zhishuang Wang**
Tianjin Key Laboratory of
    Intelligence Computing and Novel
    Software Technology
and
Key Laboratory of Computer Vision
    and System (Ministry of
    Education)
Tianjin University of Technology
Tianjin, P. R. China

**Radboud Winkels**
Leibniz Center for Law
University of Amsterdam
Amsterdam, The Netherlands

**Chengyi Xia**
Tianjin Key Laboratory of
    Intelligence Computing and Novel
    Software Technology
and
Key Laboratory of Computer Vision
    and System (Ministry of
    Education)
Tianjin University of Technology
Tianjin, P. R. China

**Yasser Yasami**
Faculty of Computer Science and
    Engineering
Shahid Beheshti University
and
Department of Computer
    Engineering and Information
    Technology
Payame Noor University
Tehran, Iran

**Chunyan Zhang**
Department of Automation
Nankai University
Tianjin, P. R. China

**Jianlei Zhang**
Department of Automation
Nankai University
Tianjin, P. R. China

**Chunyun Zhen**
Tianjin Key Laboratory of
    Intelligence Computing and Novel
    Software Technology
and
Key Laboratory of Computer Vision
    and System (Ministry of
    Education)
Tianjin University of Technology
Tianjin, P. R. China

# Chapter 1

# The Spread of Strategies Predicted by Artificial Intelligence in Networks

Chunyan Zhang

## CONTENTS

## 1.1   Individual Intelligence in Collective Behaviors

Recently, the individual and collective behaviors in multi-agent systems have received much attention of researchers from many areas, including engineering, biology, sociology, and so on. Among the related topics, the competition among strategies which indicate different benefits for the players has become the focus.

In fact, cooperation among uncorrelated individuals is necessary in order to allow the common group to offer significant advantages for them. The basic conflicts lie on the fact that, in general, the involved individuals are self-centered according to the definition. However, many examples can verify the existence of cooperative behavior in nature. For example, animals collaborate in families to raise their offspring, or in foraging groups to prey or to defend against predators. In social society, cooperation among unrelated agents will be beneficial for raising a more advanced society. For example, the altruistic cooperation among members in society can help to achieve shared goals and making efficient use of the common resource [1].

Nowadays, a growing number of researchers take an interest in evolutionary game theory, as it provides an effective framework for describing the strategy competition in several scenarios. Different with the traditional model with the assumption of full rational players and complete knowledge about the game, evolutionary models assume that agents can choose their strategies by a trial-and-error learning process. In this case, they can gradually find the strategies with better performance than others. Through this type of strategy updating in repeated games, strategies with worse performance tend to be weeded out in the system.

In this sense, as the objects of the study, the characteristics of the involved agents deserves significant attention, as the reasonable modeling about the individuals will help us to approximate to the real systems. Only in this way, the collective behaviors based on them can get better understanding. It is understandable that agents involved in the strategy competition and collective actions have (simple or complicated, the extent may be heterogeneous) intelligence. According to the introduction in artificial intelligence, an intelligent agent can be seen as an autonomous entity. She observes the variation in her surroundings and makes her action toward achieving goals. To achieve their goals, agents with intelligence may also learn or use knowledge about others or surroundings. The described intelligent agents here are closely related to agents in economics, engineering, sociology, as well as in many interdisciplinary modeling and computer simulations.

Thus based on our understanding, the hypothesis of simple agents should be relaxed, as the studied are the real agents involved in strategy competition in real social systems. Individuals here may present their intelligence in the following subjects, but observing real systems will prove that there will be more.

- *Diversity of strategy choices:* In real scenarios, agents often face multiple choices in the collective actions due to the internal and external aspects of real social systems. Different with the monotonous two-strategy profile, a set of multiple strategies can better describe the individual intelligence who decides the diversity of choices. Depending on the involved collective dilemmas, different profiles of strategy choices may be required. For example, in the collective action which establishes both temptation and punishment to defectors, strategy profile in the form of (cooperator, defector, insured players, punisher, loners, tit-for-tat, win-stay-lose-shift, etc.) can be considered [2].

- *Decision-making process:* Further, the most frequently employed setup entails that initially each player is designated either as a cooperator or defector with equal probability. Then, evolution of the two strategies is then performed in accordance with diverse update rules, such as the Fermi function [3,4], richest-following (or "learning from the best") rule [5–7], and other microscopic strategy adoption rules, such as the win-stay-lose-shift rule where the focal player has restricted information on his/her neighbors [8].

Based on our observation about real social systems, intelligent agents often have complicated emotion or social skills which play significant roles in their actions. First, real agents usually have some intelligence to predict the actions of others through their actions and current states. This type of ability can be helpful for the agents to make better decisions. To model or better describe the subsistent intelligence, the knowledge from game theory, decision theory, artificial intelligence, and so on, can make their contributions. In this sense, studying the collective behaviors in multi-agent system is an interdisciplinary topic. With the full aid of knowledge from multiple subjects, we can better explore the root causes of the happening of the collective behavior problems.

• *Artificial intelligence:* The intelligence of individuals playing games has been widely studied from many viewpoints and in different contexts. The adaptivity or self-adjusting can be seen a breakthrough point of the study. A handful of works have made their contributions about this topic. The work [9] investigates a system of adaptive agents. The players join the interated prisoner's dilemma game (PDG) governed by Pavlovian strategies in a two-dimensional spatial setting. The involved agents play with one of their neighbors sequentially and the two players update their strategies using fuzzy logic. Here, the fuzzy logic will be more appropriate to evaluate an imprecise concept like "success" than binary logic. The steady states are presented in the form of different degrees of cooperation, "economic geographies," and "efficiencies" which are related with the measures of success. Results show that the combination of spatial structure and different measure of success lead to a great diversity of statistical and spatial patterns. Though the author opted for simplicity and neglected some subtleties, they have successfully explore the combined effect of territoriality and multiple success criteria in the interaction between agents involved in spatial games.

Ishida and Katsumata [10] focuses on the generosity, that is, how many defections are tolerated. A temporal strategy involving temporal generosity, such as Tit for Tat (TFT), exhibited good performance such as noise tolerance. This work suggests that a spatial strategy with spatial generosity can maintain a cluster of cooperators by forming a membrane that protects the invasion from defectors. Here, the formation of membrane can be formulated with the spatial generosity which exceeds some threshold determined by the number of neighborhoods.

From a different point of view, Ohdaira and Terano [11] studies the role of temperate acquisitiveness in decision-making. They have reached the conclusion that the exclusion of the best decision had a significant influence on reaching an almost cooperative state. Then, they propose the second-best decision to advance the gained decision in former research. The second-best decision can be seen as the pursuit of moderate payoff and not maximum profit has some similarity with the mindset of collusive bidding. If the proposed strategy is adopted, a high level cooperative state in the PDG will be realized. In addition, the applicability of our model to the problem in the real world is discussed.

Xianyu [12] analyzes the strategy evolution in the ultimatum game of complicatedly interacting players when a number of agents display social preference. Here, three forms of agents' social preference is modeled: fairness consideration or maintaining a minimum acceptable money level, inequality aversion, and social welfare preference. Moreover, agents own incomplete information about other agents' strategies, so they need to learn and develop their own strategies in the unknown environment. Genetic Algorithm Learning Classifier System algorithm can be used to model the agents' learning issue. Results show that raising the minimum acceptable level or including fairness consideration in a game does not always promote the fairness level in the spatial ultimatum games. For the high level of minimum acceptable money and not uniform distribution of social preferences, a considerable low fairness level will be seen. And, inequality aversion exerts a negligible influence on the strategy competition, while social welfare preference facilitates the fairness level in the ultimatum game. To model potential forms of social preferences deserves to receive negligible attention in future study.

Illustrated by the abovementioned studies, the intelligence of agents can be considered to be an important factor. Therefore, it may be imperative to integrate this factor into the cooperative dilemma game through an agent-based approach. Unsolved topics are related with the questions such as: will a decision-maker always imitate their neighbors, or sometimes? A sufficient consideration to individual intelligence would be beneficial for explaining the strategy choices in social systems, which is expected in the future study? Do players choose only among the highest-scoring alternatives, or use something like proportional weighting?

The main contribution of the chapter is showing how the artificial intelligence will influence the spread of strategies in the game playing. Generally, agents are over-simplified in the traditional agent-model approaches, where the personal or social factors are overlooked. To better understand the collective behaviors of real social systems, the modeling about the agents and system should be established in a more realistic way. This chapter summarizes some works which pay attention on the individual intelligence and its application in the game playing.

## 1.2 Strategy Competition Driven by Particle Swarm Optimization

Here we make an introduction about the application of the particle swarm optimization (PSO) in strategy updating of agents who are involved in the evolutionary social dilemma. As a typical intelligent optimization algorithm, the PSO algorithm has close relation with the swarming theory [13–15]. It is well known that a physical analogy of PSO can be seen as a swarm of bees looking for a food source. In this analogy, each bee (abstracted as a particle here) uses the information stored by her memory, as well as the knowledge gained from the swarm as a whole to find the best available food source.

According to the settings in PSO, a number of simple particles are placed in the search space of some problem or function. Each particle in PSO is also endowed with a velocity vector, which plays roles in updating the current position of the particle in the swarm. Each particle evaluates the objective function at her current location, and then makes decision about her next action. The movement through the search space is updated by combining the following factors: the history of her own and the best locations with those of one or more members of the swarm, with some random perturbations.

Thus, the process synthetically makes use of the memory of each particle, and the knowledge gained by the swarm as a whole. The next iteration happens after all particles have been moved. Finally the swarm as a whole probably moves close to an optimum of the fitness function. This has some similarity with a flock of birds which are collectively foraging for food. Thus, the proposed particles have a tendency to fly toward the better search area during the search process.

Here, this optimization algorithm will be applied in the strategy updating process in the PDG. In this sense, the mentioned particle will denote a player here. Initially, a variable named as cooperation probability is endowed to each individual, and it takes value in the unit interval. Then, each player makes use of her memory (storing the information about the most profitable strategy in previous rounds), as well as the knowledge gained by the swarm (i.e., the population) to find the available one. Also, PSO makes use of a velocity vector for strategy switching of each player in the swarm. The outline is as follows:

1. An initial strategy distribution (i.e., valued by cooperation probabilities) is randomly distributed in the range $[0, 1]$;

2. Calculating the corresponding velocity vector for each strategy in the swarm;

3. Based on her previous value and the updated velocity vector, each agent updates her strategy;

4. Go to step 2 and repeat until convergence.

## 1.2.1   Model settings

The PDG with mixed-strategies is employed here. Players will be put on a square lattice of size $100 \times 100$ with periodic boundary conditions. Agents will play the iterative PDG with the neighbors on the underlying network. During each game round, players can choose from a large set of strategies, concretely denoted by the cooperation probability which values in $[0, 1]$. By using the same strategy, the focal agents play games with the neighboring partners. The simulations are started from a random strategy distribution (i.e., cooperation probabilities) in the system. The payoffs can be calculated by following the provided payoff matrix,

$$\begin{array}{cc} & \begin{array}{cc} C & \qquad\qquad D \end{array} \\ \begin{array}{c} C \\ D \end{array} & \begin{pmatrix} S_c(i) * S_c(j) & 0 \\ b * S_c(j) * (1 - S_c(i)) & 0 \end{pmatrix} \end{array}$$

where $S_c(i)$ and $S_c(j)$ represent the cooperation probabilities of participant $i$ and $j$, respectively. Based on the given payoff matrix, the two free parameters $b$ and $S_c$ here can still preserve the essence of the PDG. The positive cooperation probability means helping others at a cost to themselves, thus the cooperation probability of defectors is 0. For self-serving players, defection will be the optimal strategy for them.

Then, the spatial PDG is repeatedly played forward by following a synchronous Monte Carlo update setting. After playing the game, each player accumulates her payoffs for the usage of strategy updating. Then, agents make decision about the strategy switching for next game round. In the settings of the PSO algorithm driving the strategy update process here, all agents are endowed with the same velocity and random directions initially. At each time step, the velocity vector of $V_{i,n+1}$ of each agent $i$, will be updated in the form of

$$V_{i,n+1} = V_{i,n} + \omega(S_c(i, h) - S_c(i, n)) + (1 - \omega)(S_c(i, l) - S_c(i, n)) \qquad (1.1)$$

$$S_c(i, n + 1) = S_c(i, n) + V_{i,n+1}. \qquad (1.2)$$

where $S_c(i, h)$ is employed to denote the cooperation probability which brings the largest payoffs for her in the history.

The symbol $l$ denotes the most successful individual measured by payoffs in the system. $V_{i,n+1}$ represents the rate of the position change (velocity) for agent $i$ in game round $n + 1$. The parameter $\omega \in [0, 1]$ is the tendency between the most profitable strategy in $i$'s history or the one adopted by the most successful player $l$. Specifically, $\omega = 1$ describes the situation where the focal agent will adopt the strategy bringing the largest payoff for her in history. While, $\omega = 0$ describes the situation that the player will imitate the strategy of her neighbor gaining the largest payoff (provided it is larger than her own). Intermediate values of $\omega$ will provide chances for one going from learning the best performer ($\omega = 0$) to adopting the best action for $\omega = 1$. $P_c(i) = P_c(j) = 0$ for $\forall$ $i, j$ will be reset after each such iteration cycle.

### 1.2.2 Simulation study

The established simulations and results are summarized as follows. Here, the used network contains $10^4$ nodes and average degree 4. Initially, each of the $10^4$ agents take a cooperation probability randomly valued in $[0, 1]$. Then, the simulation proceeds until a stationary state of the system is reached. The displayed results are averaged over the last $10^4$ generations of the entire $10^5$ generations to warrant accuracy. Moreover, the gaming process was repeated for 100 realizations for each set of parameters, provided long enough times are considered.

The average cooperation probability driven by the variation of $b$ and $\omega$ is shown in Figure 1.1. Notably, average cooperation probability at the equilibrium state decreases monotonically with increasing $b$. Then, the employed PSO here significantly influence the strategy competition outcomes over the wide range of $b$. The steady cooperation probability is larger than $1/2$ for $b$ which is a little larger than 1. In $b = 1$, where defection has no advantage compared to cooperation, the cooperation evolves to a dominate state by the aid of a low $\omega$.

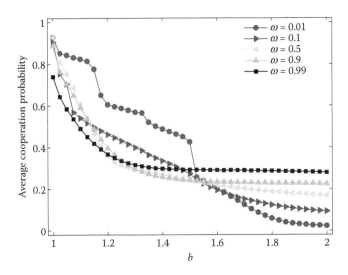

**FIGURE 1.1:** Average cooperation probability of the system in dependence on the temptation to defect $b$ for various $\omega$ values as indicated (from top to bottom $\omega = 0.01$, $0.1$, $0.5$, $0.9$, and $0.99$, respectively). Each data point of the curves results from 100 different realizations of networks to warrant appropriate accuracy. Lines connecting the symbols are just to guide the eye.

The variation of parameter $b$ will lead to two cases: $b < 1.5$ will result in a moderate dependence of the average cooperation probability on $\omega$ and $b > 1.5$ will produce a fact: the average cooperation probability is more dependent on the variation of $\omega$. Moreover, low $\omega$ (e.g., $\omega = 0.01$) promotes the spread of cooperation for small $b$, and up to $b \simeq 1.2$. However, high $\omega$ can better promote cooperation in strict cooperative dilemma.

Next, the role of $\omega$ in influencing the final strategy distribution of the population is depicted in Figure 1.2. At the case of $\omega = 0.01$, the strategy distribution is monotonous. However, for large values of $\omega = 0.99$, the strategy distribution presents diversity. When $\omega = 0.01$, the learning from successful neighbors will facilitate the spreading of strategies with better performance. As shown in Figure 1.2a, the steady strategy distribution of the system shows polarization, where $P_c = 1$ or $P_c = 0$. Otherwise, for sufficiently large $\omega$ (e.g., $\omega = 0.99$), the strategy updating of individuals is based on their best actions in their history.

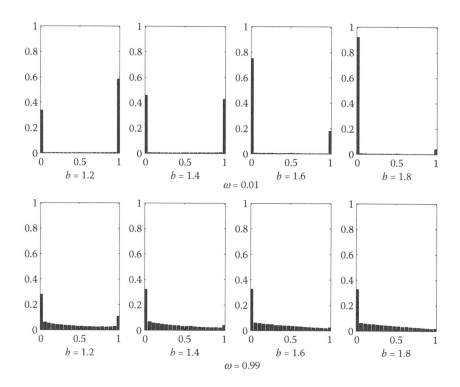

**FIGURE 1.2:** Dependence of the proportion distribution of cooperation probabilities on the defection temptation $b$. Top row depicts results for the $\omega = 0.01$, while the bottom row features results for $\omega = 0.99$. Note that the horizontal axis displays the value of the cooperation probability, while the vertical axis is the proportion of individuals with such cooperation probability in the whole population. The employed system size was $N = 10^4$.

In this case, probably different information from respective history will lead to the strong diversity of strategy distribution in the system. Further confirmation about the influence of the larger $\omega$ in promoting cooperation spreading, when the degree of collective dilemma turns rigorous from mild, is shown in Figure 1.1.

Figure 1.3 provides some snapshots about the cooperation probability distribution in the system. $\omega = 0.01$ will lead to the isolation of homogeneous groups of players with cooperation probability $P_c = 0$ and $P_c = 1$ respectively, as seen in the upper panels of Figure 1.3. Then, the heterogeneous kaleidoscope shown in the bottom panel of Figure 1.3 suggests the emergence of many clusters

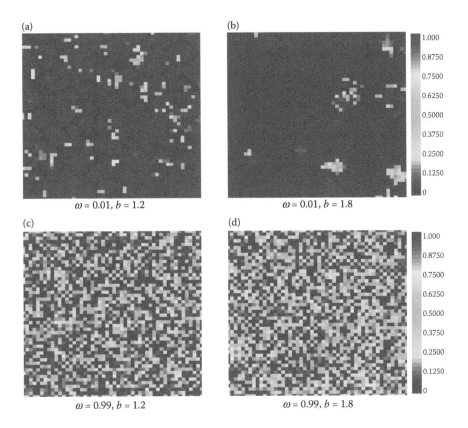

**FIGURE 1.3:** Characteristic equilibrium spatial distributions of cooperation probabilities of the whole population in dependence on several values of $\omega$ and $b$, respectively. Top row depicts results for the $\omega = 0.01$, while the bottom row features results for $\omega = 0.99$. In both rows, the game parameter $b$ is 1.2 and 1.8 from left to right. All panels are a $50 \times 50$ portion of the full $100 \times 100$ spatial grids at a certain time when the system reaches a steady state. The color code indicates the value of cooperation probability. Each data point on the lattice denotes a player, and the color on the point correspondingly expresses as the cooperation probability of the player. For more details of the results with color, please see [35].

consisted of players with higher cooperation probabilities. In this case, the strategy with better performance spread slowly among the system. Actually, such type of clusters is beneficial for the players with higher cooperation probabilities to spread their strategies. The high cooperation level within clusters will help the involved agents to resist the invasion of players with low cooperation probability. In fact, players in such clusters can collect the benefits of mutual cooperation, with which to resist the exploitation of players with low cooperation probability along the boundaries of clusters.

Then, the individual velocity of the population at the steady state is another focus. Results summarized in Figure 1.4 show that a majority of the individual velocities are not 0, even though no significant change can be seen from the average cooperation probability. As mentioned, the values of $\omega$ can significantly affect the strategy competition result. For small $\omega$ (e.g., $\omega = 0.01$), a high heterogeneity of the individual velocity of the population can be observed from the corresponding panels in Figure 1.4). However, for relatively large $\omega$, for example, $\omega = 0.99$, the individual velocity of the system show small fluctuations which are close to the value of 0 (Figure 1.5).

*A brief summarization:* To closely model the individual intelligence in their decision-making process, the influence of PSO in the spatial two-player-mixed-strategies PDG is introduced here. Individuals update their strategies by following the rules of the PSO algorithm. Especially, each player makes use of the most profitable strategy in her memory and the best available one in the swarm. Results suggest that imitating the most profitable strategy in the population would benefit the survival of cooperation especially when the degree of collective dilemma turns rigorous from mild. The steady strategy distribution of the system is closely related with the parameter determining the strategy update. In this case, how to control the strategy spreading can get some inspiration from the combination of intelligent algorithm and evolutionary game theory here.

## 1.3   Strategy Spreading Controlled by Fuzzy Neural Network

- *Artificial neural network:* The biologically inspired intelligence algorithm of artificial neural networks is now put into used in many fields. It can be seen as a viable, multipurpose, robust computational method with the solid theoretic support. Moreover, it also has strong potential applications in many areas. For example, artificial neural networks can fetch new medical information from the gained raw data, and then establish models that are beneficial for medical decision-making. Nowadays, many applications by employing the artificial neural network are emerging, and more anticipated results are on the way [16].

  Specifically, artificial neural networks can be seen as a computational approach that establishes multifactorial analysis [17]. As an interdisciplinary topic, it is inspired by the networks of biological neurons in

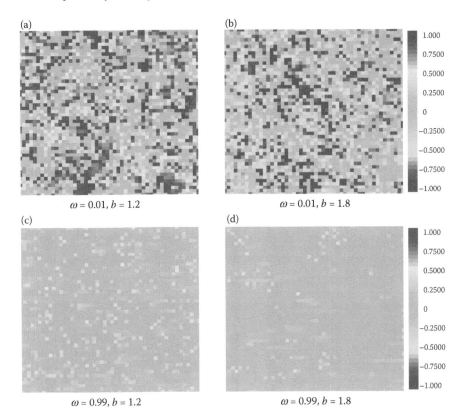

**FIGURE 1.4:** Typical snapshots of individual velocity in the whole population in dependence on different values of $\omega$ and $b$. Top row depicts results for the $\omega = 0.01$, while the bottom row features results for $\omega = 0.99$. In both rows, the game parameter $b$ is 1.2 and 1.8 from left to right. The size of the network is $N = 10^4$ nodes. All snapshots are a $50 \times 50$ portion of the full $100 \times 100$ spatial grids at a certain time when the system reaches a steady state. The color code indicates the value of individual velocity. The snapshots are contractible picture of large image scale of the system on the full $100 \times 100$ spatial grids at a certain time when the system reaches a steady state. For more details of the results with color, please see [35].

humans. In general, it contains a set of simple nonlinear computing elements whose inputs and outputs are tried together to form a network. According to the settings, there are many layers of simple computing nodes that operate as nonlinear summing devices in the artificial neural network [17]. These nodes are interconnected by weighted connection lines, and the values of weights can be adjusted when data are presented to the network during the training process. After the successful training, artificial neural network can be used for predicting, classifying,

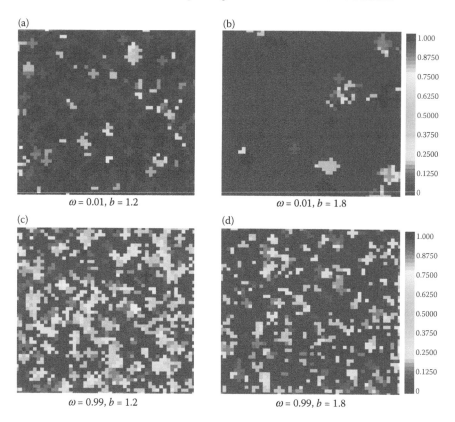

(a)    $\omega = 0.01, b = 1.2$

(b)    $\omega = 0.01, b = 1.8$

(c)    $\omega = 0.99, b = 1.2$

(d)    $\omega = 0.99, b = 1.8$

**FIGURE 1.5:** Characteristic snapshots of cooperation probability in the whole population in dependence on different values of $\omega$ and $\eta$. Top row depicts results for the $\omega = 0.01$, while the bottom row features results for $\omega = 0.99$. In both rows, the game parameter $b$ is 1.2 and 1.8 from left to right. All panels a 50 × 50 portion of the full 100 × 100 spatial grids at a certain time when the system reaches a steady state. The color code indicates the value of individual cooperation probability. For more details of the results with color, please see [35].

approximating a function, recognizing a pattern in multifactorial data, and so on [18].

The first computational, trainable neural networks were developed with two layers of computational nodes and a single layer of interconnections [19]. The shortage is that this model was limited to the solution of linear problems. The past decades have witnessed a variety of neural network paradigms have been developed, analyzed, studied, and applied in the wider range of areas [20,21]. The applications include but are not limited to: the automatic target recognition, control of flying aircraft, and fault detection in complex engineering systems, time series predictions, decision support roles in the financial industry.

- *Fuzzy neural network:* Artificial neural networks can only come into play if the problem to be solved is expressed by a sufficient number of observed examples. These observations are used to train the black box. Here, no prior knowledge about the unsolved problem needs to be given, and it is not straightforward to extract comprehensible rules from the artificial neural network's structure.

A fuzzy system requires the linguistic rules rather than learning examples as prior knowledge. And in the settings of fuzzy system, the involved input and output variables need to be described linguistically. In the presence of incomplete, wrong or contradictory knowledge, the fuzzy system needs to be tuned. Owing to the lack of a formal approach for it, the tuning has to be established in a heuristic way, where time consuming and error-prone will be unavoidable. Briefly, the traditional fuzzy system is closely related with the experts' knowledge, which may be not very objective. And, it is not easy to gain the robust knowledge and available human experts. Recently, the learning algorithm in the framework of artificial neural networks has been applied to improve the performance of a fuzzy system, which is expected to be a promising way.

The artificial neural networks and fuzzy systems share the common characteristics that: the possibility of being applied in the solution of problems, for example, pattern recognition, regression, or density estimation, if mathematical model of the given problem is not available. Their respective shortcomings will almost completely disappear by combining both concepts.

In this sense, fuzzy neural network (FNN) control systems have received extensive attention. For example, the work [22] introduces a general neural network model for a fuzzy logic control and decision system. The proposed model is trained for controlling an unmanned vehicle in the way of combining unsupervised and supervised learning. Horikawa [23] proposes a FNN where the expert control rules are learned, and in Reference 24, the adaptive neurons combined with a fuzzy logic controller to provide solutions for the pole balancing problem. However, in all of these systems a teacher responsible for training is required. Furthermore, adaptation to changes in the environment is not provided for.

## 1.3.1 Problem statement

A typical assumption is that players always interact with all of their neighbors with sufficient interaction strength during the game. From the perspective of real social systems, this ideal assumption needs to be relaxed. In the work of Traulsen et al. [25], each pair of individuals in finite populations interact with a probability, resulting a uniform distribution of interaction numbers. Then, Chen et al. [26] studied the random interaction through a fixed intensity of interaction in spatial repeated PDG, and an optimal region of interaction strength resulting in maximum cooperation level can be found.

It is well known that reputation can help individuals recognize *good and bad guys*, acting as a significant factor to help individuals adjust their partnerships or carry out the selective interactions [27–29]. To approach the reality, a continuous variable of evaluation level proposed by the work [30] can more accurately describe one's behavior. In the spatial PDG, they assume that each player engages in pairwise interaction based on an intensity of interaction $W_{x,y}(t)$. $W_{x,y}(t)$ denotes the possibility that player $x$ plays the game with her neighbor $y$ at time $t$. Larger $W_{x,y}(t)$ indicates a larger likelihood of interaction. Here, the variation of $W_{x,y}(t)$ is closely related with the individual reputation.

- *Prisoner's dilemma game:* The typical model of PDG is used here to describe the interest conflicts between the individual and the group. Here, $s_x$ is used as the strategy of player $x$. And it follows two simple strategies, $s_x(t) = [1, 0]^T$ corresponds to cooperation ($C$), and $s_x(t) = [0, 1]^T$ corresponds to defection ($D$) at time $t$ step.

- *Formation of the interaction intensity:* For a pair of players $x$ and $y$, $w_{x \to y}(t)$ $[w_{y \to x}(t)]$ denotes the probability of interaction from player $x(y)$ to player $y(x)$ at time $t$. Initially, $w_{x \to y}(t)$ $[w_{y \to x}(t)]$ randomly takes values in the interval $(0, 1)$. Owing to the independence of unilateral intention interaction, the initial interaction intensity between $x$ and $y$ is

$$W_{x,y}(0) = w_{x \to y}(0) \cdot w_{y \to x}(0). \tag{1.3}$$

The interaction intensity at time $t(t \geq 1)$ will be

$$W_{x,y}(t) = w_{x \to y}(t) \cdot w_{y \to x}(t). \tag{1.4}$$

After interaction, player $x(y)$ probably unilaterally revises the intensity of interaction from $x(y)$ to $y(x)$, thus resulting the variation $\Delta w_x$ $(\Delta w_y)$. Accordingly, at time $t+1$ the interaction intensity between them is as follows:

$$
\begin{aligned}
W_{x,y}(t = 1) &= w_{x \to y}(t = 1) \cdot w_{y \to x}(t + 1) \\
&= [w_{x \to y}(t) + \Delta w_x] \cdot [w_{y \to x}(t) - \Delta w_y]. \tag{1.5}
\end{aligned}
$$

where $\Delta w_x$ is increment and $\Delta w_y$ is the decrement

- *Reputation of players*

  **Definition 1.1. Effective neighbor:** *In the process of random interaction, an individual $x$ interact with her neighbors according to their interaction intensity. Thus, individual $x$ may finally interact with part of her* neighbors. Thus, the neighbors who finally join in the game with $x$ can be seen as the effective neighbors of $x$.

Then player $x$'s reputation at time $t$ is determined by

$$R_x(t) = 0.5 R_x(t-1) + 0.5 \operatorname{sgn}\left[s_x(t)\right] \frac{N_x(t)}{k_x}\phi, \qquad (1.6)$$

where

$$\operatorname{sgn}[s_x(t)] = \begin{cases} 1 & s_x(t) = C \\ -1 & s_x(r) = D \end{cases}, \quad 0 \le \phi \le 1$$

Here, $\phi$ denotes the evaluation level about individuals. For $\phi = 0$, the player $x$'s reputation gradually varies from the initial value and approach zero as game proceeds. According to the rules (6) (in what follows), this kind of change of reputation simply do not work anymore for the adjustment of the interaction intensity. In other words, it will transform to the classical random interaction. For $0 < \phi < 1$, larger $\phi$ indicates the higher distinguishing extent for player $x$'s behavior. For $\phi = 1$, the interaction partners of player $x$ unexpectedly objectively evaluate her behavior, whether she is good or bad. $N_x(t)$ means the number of player $x$'s effective neighbors at time $t$, and $k_x$ denotes the number of all $x$'s neighbors.

## 1.3.2    Framework of model

The iterated PDG is performed on a $L \times L$ square lattice with periodic boundary conditions. Each player engages in pairwise interactions within her von Neumann neighborhood. The nodes of dynamical graph represent players, the edges denotes the pairwise partnership, and their weights denote their interaction intensity respectively. The rescaled payoff matrix $M$ depends on one single parameter $b$: $T = b > 1$, $R = 1$, and $P = S = 0$.

$$\begin{array}{cc} & \begin{array}{cc} C & D \end{array} \\ \begin{array}{c} C \\ D \end{array} & \begin{pmatrix} 1 & 0 \\ b & 0 \end{pmatrix} \end{array}$$

The evolutionary process will thus be

---

**Stage 1: Stochastic interaction**

Each player $x$ engages in the pairwise interactions with the interaction intensity $W_{x,y}(t)$, and using the same strategy $s_x(t)$ within her neighbors.

---

The collected payoff will be

$$U_x(t) = \sum_{y \in \Omega_x(t)} s_x^T(t) M s_y(t), \qquad (1.7)$$

where $\Omega_x(t)$ represents the set of effective neighbors of player $x$, and $M$ is the payoff matrix.

### Stage 2: Updating of the intensity of interaction

After playing the game, each individual's reputation and interaction intensity will be updated. Whether the interaction intensity from player $x(y)$ to her partner $y(x)$ will unilaterally change is determined by player $y$'s ($x$'s) reputation. At time $t$, if player $y$ has a good reputation $[R_y(t) > 0]$, and $[R_y(t) > R_y(t-1)]$, $x$ will unilaterally increase the interaction intensity with $y$ in the form of $[\Delta R_y(t-1, t) = R_y(t) - R_y(t-1)]$. However, player $x$ will unilaterally reduce the interaction intensity with $y$ due to the bad reputation of player $y(R_y(t) < 0)$, and $[R_y(t) < R_y(t-1)]$. At time $t+1$, the interaction intensity $[w_{x \to y}(t+1)$ is

$$\begin{cases} w_{x \to y}(t) + \Delta w_x & \text{if } R_y(t) > 0 \quad \text{and} \quad R_y(t) > R_y(t-1) \\ w_{x \to y}(t) - \Delta w_x & \text{if } R_y(t) < 0 \quad \text{and} \quad R_y(t) < R_y(t-1) \end{cases} \qquad (1.8)$$

In other cases, the interaction intensity remains fixed.

Player $y$ will unilaterally change the intensity of interaction with $x$ in the similar way. To ensure the positive values of the interaction intensity, its smallest value is set to 0.0001, and the maximum value will be 1.

### Stage 3: Strategy switching

Individuals have the chance to update their strategy, and the Fermi rule [31] is adopted here. Player $x$ randomly selects a neighbor $y$ and imitates $y$'s strategy with the probability determined by the total payoff difference between them,

$$f(U_y - U_x) = \frac{1}{1 + \exp\left[-(U_y - U_x)/K\right]}, \qquad (1.9)$$

where $K$ quantifies the uncertainty that may occurs in the decision-making process.

For simplicity, we assume that each individual has the same evaluation level, and mainly investigate how the evaluation level ($\phi$) affects the strategy competition in spatial PDG.

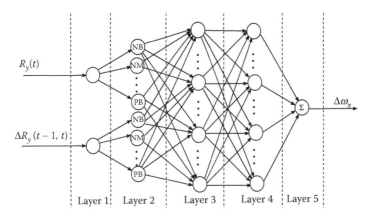

**FIGURE 1.6:** Schematic diagram about the used FNN. Player $x$ adjusts the interaction intensity with her neighbor $y$ base on $y$'s reputation. Layer 1 is input layer, followed by the membership layer, the fuzzy rules layer, the normalized processing layer, and the output layer. The neuronic number of each layer is 2, 12, 36, 36, and 1, respectively.

*Rules of adjusting $\Delta w_x$:* The FNN [32,33] is used to adaptively adjust the interaction intensity. Fuzzy language has a certain vagueness. For example, an individual's reputation can be described by one of these terms such as worst, worse, bad, good, or best, which share some characteristic of fuzziness. Therefore, the reputation $R_y(t)$ of player $y$ and the variable quantity $\Delta R_y(t-1, t)$ of the reputation can be seen as the input vector of FNN. Then the variable quantity $\Delta w_x$ of the interaction intensity is taken as the output of FNN. In addition, in FNN, the interaction intensity takes values in $\Delta w_x \in [-0.05, 0.05]$ to realize a gradual and not significant variation. As shown in Figure 1.6, the FNN here consists of five layers.

---

**Layer 1: Input layer**

Each node of input layer, respectively, connects with each component of input vector $R_i$, and the total number of nodes (denote by $N_1$) of this layer equal to the dimension of input vector. Notably, the nodes in this layer just transmit input signals to the next layer directly, that is,

$$I_i^{(1)} = R_i, \quad O_i^1 = I_i^{(1)}. \quad (i = 1, 2, \dots, N_1) \qquad (1.10)$$

Obviously, here $N_1 = 2$,

$$\begin{bmatrix} R_y(t) \\ \Delta R_y(t-1, t) \end{bmatrix} = \begin{bmatrix} R_y(t) \\ R_y(t) - R_y(t-1) \end{bmatrix} \qquad (1.11)$$

---

## Layer 2: Membership layer

In this layer, a single node can be seen as a membership function, thus its output is the degree of membership, suggesting a signal component which is subject to the extent of corresponding fuzzy language set. Here, a bell-shaped function is chosen,

$$I_{ij}^{(2)} = -\frac{(x_i - a_{ij})^2}{b_{ij}^2}, \quad O_{ij}^{(2)} = A_{ij} = \exp(I_{ij}^{(2)}), \quad (1.12)$$

where $i = 1, 2, \ldots, N_1$; $j = 1, 2, \ldots, m_i$; $a_{ij}$ and $b_{ij}$ are, respectively, the center and the width of the bell-shaped function of the $j$th term of the $i$th input variable $R_i$. The number of all the nodes of this layer $N_2 = \sum_{i=1}^{n} m_i$, where $m_i$ denotes the number of fuzzy subsets of signal component. The reputation of the player $y$ at time $t$ $R_y(t) = R_1 \in [-1,1]$, and the variation $\Delta R_y(t-1, t) = R_2 \in [\min\{R_y(t)\} - \max\{R_y(t-1)\}, \max\{R_y(t)\} - \min\{R_y(t-1)\}] = [-2,2]$. The interval of each input component $R_i(i = 1,2)$ is divided into six fuzzy classes, as shown in Figure 1.7. Based on this consideration, the number of membership functions $m_1 = m_2 = C_2^1 C_3^1 = 6$, and there are 12 nodes in this layer ($N_2 = m_1 + m_2 = 12$).

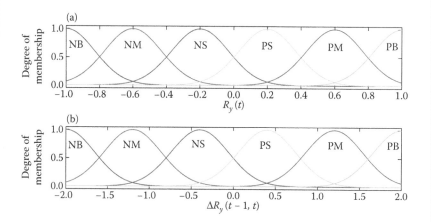

**FIGURE 1.7:** The distribution of fuzzy subsets' membership degree for each input component. (a) the reputation $R_y(t) \in [-1, 1]$ of player $y$; (b) the variation $\Delta R_y(t-1, t) \in [-2, 2]$ of player $y$'s reputation. The interval of each input will be evenly divided into six fuzzts, denoted by six variables (i.e., NB, NM, NS, PS, PM, and PB).

### Layer 3: Fuzzy rules layer

Every node represents a fuzzy rule, and the rule node perform the fuzzy because the links in this layer are used to perform precondition matching of fuzzy logic rules. Thus, the number of nodes in this layer is $N_3 = m = \prod_{i=1}^{N_1} m_{N_1}$, the fitness of fuzzy rule

$$\alpha_l = \min \{A_{1j_1}, A_{2j_2}, \ldots, A_{N_1 j_{N_1}}\}, \tag{1.13}$$

where $j_{N_1} \in \{1, 2, \ldots, m_{N_1}\}$, $j_1 \in \{1, 2, \ldots, 6\}$, $j_2 \in \{1, 2, \ldots, 6\}$. Correspondingly,

$$I_l^{(3)} = \alpha_l, \quad O_l^{(3)} = I_l^{(3)}. \tag{1.14}$$

### Layer 4: Normalized processing layer

The number of nodes of this layer is the same with that in layer 3. The function of this layer is the normalized calculation,

$$\bar{\alpha}_l = \frac{\alpha_l}{\sum_{i=1}^{m} \alpha_i}. \quad (l = 1, 2, \ldots, m) \tag{1.15}$$

### Layer 5: Output layer

The function of this layer is the defuzzification, and there is just one node, $N_5 = 1$.

$$\Delta w_x = \sum_{j=1}^{m} v_{ij} \bar{\alpha}_j, \quad (i = 1, 2, \ldots, r) \tag{1.16}$$

where $r$ denotes the number of output nodes and $v_{ij}$ represents the weight values between layer 4 and layer 5.

Next, the learning algorithm of FNN and the methods for obtaining the necessary spatial sample will be introduced.

By the error cost function

$$E = \frac{1}{2} \sum_{i=1}^{r} (t_x - \Delta w_x)^2, \tag{1.17}$$

where $t_x$ is the value of expectation, and based on erroneous reversed dissemination method, it is true that

$$
\begin{cases}
\dfrac{\partial E}{\partial v_{ij}} = -(t_x - \Delta w_x)\bar{\alpha}_l \\[3mm]
\dfrac{\partial E}{\partial a_{ij}} = \dfrac{\partial E}{\partial I_{ij}^{(2)}} \dfrac{\partial I_{ij}^{(2)}}{\partial a_{ij}} \\[3mm]
\dfrac{\partial E}{\partial b_{ij}} = \dfrac{\partial E}{\partial I_{ij}^{(2)}} \dfrac{\partial I_{ij}^{(2)}}{\partial b_{ij}}.
\end{cases}
\tag{1.18}
$$

So, the learning algorithm of the parameter adjustment is

$$
v_{ij}(k+1) = v_{ij}(k) - \beta \frac{\partial E}{\partial v_{ij}}
\tag{1.19}
$$

$$
a_{ij}(k+1) = a_{ij}(k) - \beta \frac{\partial E}{\partial a_{ij}},
\tag{1.20}
$$

$$
b_{ij}(k+1) = b_{ij}(k) - \beta \frac{\partial E}{\partial b_{ij}},
\tag{1.21}
$$

where $\beta$ is the learning efficiency. The interval of output variable ($\Delta w_x \in [-0.05, 0.05]$) is evenly divided into six fuzzy classes before defuzzification, denoted by six variables from NB to PB. The fuzzy control rules are shown in Figure 1.8. The main parameters of membership functions ($a_{ij}$, $b_{ij}$) and the weight values ($v_{ij}$) can get self-adjustment.

| | | *and $\Delta R_y (t-1, t)$ is* | | | | | |
|---|---|---|---|---|---|---|---|
| | | **NB** | **NM** | **NS** | **PS** | **PM** | **PB** |
| | **NB** | NB | NM | NS | ⊖ | ⊖ | ⊖ |
| | **NM** | NB | NM | NS | ⊖ | ⊖ | ⊖ |
| *if $R_y(t)$ is* | **NS** | NB | NM | NS | ⊖ | ⊖ | ⊖ |
| | **PS** | ⊖ | ⊖ | ⊖ | PS | PM | PB |
| | **PM** | ⊖ | ⊖ | ⊖ | PS | PM | PB |
| | **PB** | ⊖ | ⊖ | ⊖ | PS | PM | PB |

**FIGURE 1.8:** The fuzzy control rules. The interaction intensity is closely related with the reputation and its variation. For $R_y(t) > 0$, larger $\Delta R_y(t-1, t)$ will lead to higher $\Delta w_x$. While for $R_y(t) < 0$, larger $|\Delta R_y(t-1, t)|$ will lead to increasing $|\Delta w_x|$. Here, $|\Delta R_y(t-1, t)|$ denotes the absolute value of $\Delta R_y (t-1, t)$, and the same for $|\Delta w_x|$. In the case of ⊖, $\Delta R_x$ will be equal to zero.

### 1.3.3 Dynamics of strategy competition

Simulations are performed on a square lattice of size $100 \times 100$, with a random strategy distribution in the system. Player $x$'s reputation $R_x(0)$ randomly takes values in the interval $[-1, 1]$. $R_x(0) < 0$ indicates $x$'s bad reputation at time $t = 0$, while $R_x(0) > 0$ denotes the good reputation. $R_x(0) = 0$ represents a neutral reputation.

#### 1.3.3.1 Effect of evaluation level and verification

To quantify how the variation of interaction intensity influences the strategy competition, the dependence of the cooperation level $f_c$ on $\phi$ for different values of $b$ is summarized by Figure 1.9. The specific influence of $f_c$ on $\phi$ also depends on the degree of cooperative dilemma $b$. For small ($b = 1.05$) or moderate ($b = 1.10$) $b$, larger $\phi$ increases the value of $f_c$, and even its domination of the system after $\phi$ reaches some critical value. Further, increasing $b$ will shrinkage the area with high cooperation. When the value of $b$ is beyond 1.142 (e.g., $b = 1.15$), results suggest that there exists an optimal value of $\phi$ (approximately $\phi \approx 0.68$) which can best promote the spreading of cooperation. In the presence of larger $b$, appropriate assessment level will be the best for the cooperation spreading. Then for $b(b > 1.15)$, the $f_c$ will be notably weakened, and even approaches zero for large $b \rightarrow 1.175$.

#### 1.3.3.2 Microscopic behaviors of individuals

For an intuitive understanding, the microscopic behaviors of players are demonstrated in the form of typical snapshots of the system with time $t$ for $b = 1.10$ and $\phi = 0.8$ (see Figure 1.10). A random strategy distribution is

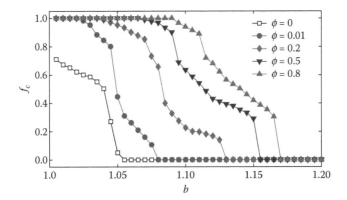

**FIGURE 1.9:** The steady cooperation level $f_c$ in dependence on the temptation to defect $b$ and $\phi$. The PDG is performed by players situating on a square lattice with periodic boundary conditions. Larger *phi* will provide a wider range $b$ for the survival of cooperators. Parameter settings here: $L = 100$, $K = 0.1$.

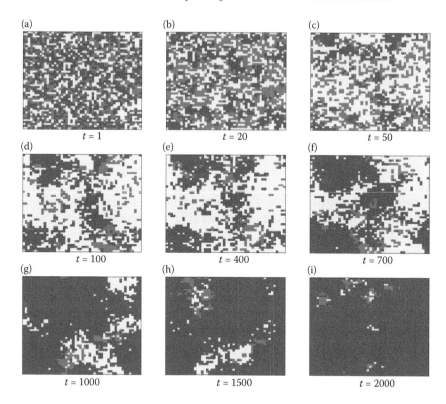

**FIGURE 1.10:** Snapshots about distributions of cooperators (blue), defectors (red), and loners (yellow) on a square lattice obtained for $b = 1.10$ and $\phi = 0.8$ at different time $t$. Results show that cooperators can get survival and spreading by the aid of forming clusters as game proceeds. (a) $t = 1$ $[f_c(1) = 0.3652]$, (b) $t = 20$ $[f_c(20) = 0.2768]$, (c) $t = 50$ $[f_c(50) = 0.1996]$, (d) $t = 100$ $[f_c(100) = 0.3072]$, (e) $t = 400$ $[f_c(400) = 0.5052]$, (f) $t = 700$ $[f_c(700) = 0.7924]$, (g) $t = 1000$ $[f_c(1000) = 0.8724]$, (h) $t = 1500$ $[f_c(1500) = 0.9064]$, and (i) $t = 2000$ $[f_c(2000) = 0.9432]$. It is worth noting that the loners here are different from the one by traditional definition, where loners will refuse to join in the game but gain a fixed benefit.

established in the system at $t = 0$. And, the interaction intensity between pairs of players randomly takes values in the interval $(0, 1)$. In this case, some loners may exist in the initialized system (shown in Figure 1.10a), due to the potential tiny interaction intensity with their partners. In the process shown in Figure 1.10b and c, defection gets spreading due to the higher payoffs, even suffering a drop in their reputation. Then individuals with poor reputation may tend to be loners due to the self-regulation of interaction intensity. However, the clustering can provide cooperators with long-term benefit, thus the decrement of $f_c$ can gradually be restrained. Driven by the proposed interaction intensity here, the gaming interactions between cooperators are steady and beneficial for the spreading of cooperation. But that could change for the pairs of defectors, where feeble interaction intensity may exist. As game proceeds, the disintegrated $D$-clusters and stronger $C$-clusters can be seen as shown in Figure 1.10f–i. Finally, the cooperation level $f_c$ of the system gets be promoted.

### 1.3.3.3   Formation of $C$-clusters

By the aid of the concept of the node $i$'s in-strength and out-strength [34], the formation $C$-clusters is further investigated. Generally for directed network, the degree of node includes out-degree and in-degree. The out-degree $k_i^{out}$ of node $i$ denotes the number of links from $i$ to her neighboring nodes. Meanwhile, in-degree $k_i^{in}$ is the number of links from $i$'s neighboring nodes to $i$. However, for weighted networks where the weight matrix is $\Theta = [w_{ij}]$, the strength of node $i$ will be

$$\theta_i = \sum_{j=1}^{k_i} w_{ij}. \tag{1.22}$$

In this case, the in-strength and out-strength of directed-weighted networks are

$$\theta_i^{in} = \sum_{j=1}^{k_i^{in}} w_{ji}, \quad \theta_i^{out} = \sum_{j=1}^{k_i^{out}} w_{ij}. \tag{1.23}$$

And the average value of in-strength $(\overline{\theta_i^{in}})$ and out-strength $(\overline{\theta_i^{out}})$ will be

$$\overline{\theta_i^{in}} = \frac{1}{k_i^{in}} \sum_{j=1}^{k_i^{in}} w_{ji}, \quad \overline{\theta_i^{out}} = \frac{1}{k_i^{out}} \sum_{j=1}^{k_i^{out}} w_{ij}. \tag{1.24}$$

**Definition 1.2 Attraction degree:** *In the evolutionary games here, the average value of a node's in-strength $(\overline{\theta_i^{in}})$ describes her attraction degree for neighboring partners.*

The conditions of $R_x(t) > 0$ and $R_x(t) > R_x(t-1)$ will increase player $x$'s attraction degree $\theta_x^{in}$, the numbers of her effective neighbors, and finally, her reputation in the next interaction. The abovementioned process will then facilitate the spreading of cooperation by the aid of $C$-clusters. Whereas the condition of $R_y(t) < 0$ and $R_x(t) < R_x(t-1)$ will decrease player $y$'s attraction degree $\theta_y^{in}$, and the number of her effective neighbors.

*A brief summarization:* In the real society, the interaction and its tensity between pairs of game players may take dynamic changes. Here, this variation is based on the reputation of players, in terms of the evaluation level $\phi$ of individual behavior. The adjustment takes some hints from the ideas of FNN. $\phi > 0$ will cause a series of reactions, such as the individual's reputation, the number of her effective neighbors, the formation of cooperator clusters, and finally, the spreading of cooperation strategy. Besides, results suggest the still enhanced cooperation in strict cooperative dilemma when the FNN is employed here.

## 1.4   Conclusion

The competition among strategies which bring different benefits for the players in real social systems is of both scientific and practical interest. Plenty of research examine the influencing factors that drive the strategy spreading. One of the limitations about the traditional agent-model approaches is that these binary and completely deterministic agents are an over-simplification of real individuals. More personal or social factors need to consider, thus to incorporate the due intelligence of real agents. This chapter summarizes some previously published work which model the individual intelligence from different points of view.

It is worth remarking that there are some other interesting points that can combine the individual intelligence and strategy competition. For example, genetic algorithms can be seen as an effective search algorithms for the optimization or classification problems. Based on the mechanics of natural selection and genetics, this algorithm works by repeatedly modifying a population of artificial structures through the application of genetic operators. For the rational agents, how to choose the strategy which can bring the largest benefits for her, can lend some hints from the optimization algorithm.

Further, a fuzzy system can be implemented to dynamically adapt the inertia weight of the PSO algorithm. The fuzzy adaptive PSO has been verified to be a promising optimization method, which is especially useful for optimization problems with a dynamic environment. In this sense, its role in influencing the strategy competition of agents involved in complicated surroundings is worth expecting.

Additionally, existing bodies of research on evolutionary games have paid their attention on the natural environments of involved agents. However, the choice of a single-layered network as the spatial structure does not seem very realistic. The underlying structure of various complex systems in everyday life (from social networks to the World Wide Web) and in nature (cell's metabolic system, epidemics, ecological networks, etc.) may be better described by complicated and multilayered networks. Hence, a future issue to address is to combine the strategy competition with networks having more realistic structures. The spatially extended games are necessary because natural surroundings is not a negligible issue to understand the strategy competition dynamics. This requires further research.

Besides, information can be seen as a significant factor in the game playing. Therefore, it may be helpful to integrate some realistic factors about information into the strategy competition as well. For example, in many cases, the player can not obtain the exact information on her opponents' information about games and is unsure of other agents' strategies. In this situation, it is not possible for the agents to simply imitate the best strategy of their neighboring agents, which remains issues worth studying.

Summarizing, depending on the particular context, more sophisticated agents should be considered in the agent model of games. For example, agents are endowed with the memory of previous encounters or with the age for game playing, and so on. Only in this way, the collective behaviors of agents living in a society can get better understanding in a more realistic way, when the agent-based model is extended to this version. Further study needs to be conducted in this field, with special focus on the effect of agents' intelligence on the strategy spreading.

## 1.5 Glossary

**Strategy competition:** In the evolutionary game theory, players can update their strategies, and that bringing higher payoffs will get spreading in the system of the self-centered agents.

**Individual intelligence:** For the agents living in a society, they may possess varying forms of intelligence which help them to make decisions in the gaming playing.

**Particle swarm optimization:** It is a computational approach that optimizes a problem by iteratively trying to improve a candidate solution with respect to a given measure of quality.

**Fuzzy neural network:** It is a learning machine that accesses the parameters of a fuzzy system by exploiting approximation from neural networks.

**Genetic algorithms:** It is a search algorithm for the optimization or classification problems.

## References

1. J. M. Smith and E. Szathmáry. *The Major Transitions in Evolution*. Oxford, UK: W. H. Freeman & Co, 1995.

2. T. Chu, J. Zhang, and F. J. Weissing. Does insurance against punishment undermine cooperation in the evolution of public goods games? *J. Theor. Biol.*, 321:78–82, 2013.

3. A. Traulsen, J. M. Pacheco, and M. A. Nowak. Pairwise comparison and selection temperature in evolutionary game dynamics. *J. Theor. Biol.*, 246:522–529, 2007.

4. F. C. Santos and J. M. Pacheco. Scale-free networks provide a unifying framework for the emergence of cooperation. *Phys. Rev. Lett.*, 95:098104, 2005.

5. G. Abramson and M. Kuperman. Social games in a social network. *Phys. Rev. E*, 63:030901(R), 2001.

6. M. G. Zimmermann, V. Eguíluz, and M. S. Miguel. Coevolution of dynamical states and interactions in dynamic networks. *Phys. Rev. E*, 69:065102(R), 2004.

7. J. Tanimoto. Promotion of cooperation through co-evolution of networks and strategy in a 2 × 2 game. *Phys. A*, 388:953–960, 2009.

8. G. Szabó and G. Fáth. Evolutionary games on graphs. *Phys. Rep.*, 446:97–216, 2007.

9. H. Fort and N. Pèrez. The fate of spatial dilemmas with different fuzzy measures of success. *J. Artif. Soc. Soc. Simul.*, 8(3), 2005. http://jasss.soc.surrey.ac.uk/8/3/1.html.

10. Y. Ishida and Y. Katsumata. A note on space-time interplay through generosity in a membrane formation with spatial prisoner's dilemma game. *Lect. Notes Artif. Intell.*, 5179:448–455, 2008.

11. T. Ohdaira and T. Terano. Cooperation in the prisoner's dilemma game based on the second-best decision. *J. Artif. Soc. Soc. Simul.*, 12(4):7, 2009.

12. B. Xianyu. Social preference, incomplete information, and the evolution of ultimatum game in the small world networks: An agent-based approach. *J. Artif. Soc. Soc. Simul.*, 13(2):7, 2010.

13. J. Kennedy et al. Particle swarm optimization. In *Proceedings of IEEE International Conference on Neural Networks*, 4, 1942–1948. Piscataway, NJ: IEEE, 1995.

14. R. C. Eberhart and J. Kennedy. A new optimizer using particle swarm theory. In *Proceedings of the Sixth International Symposium on Micro Machine and Human Science*, 43:39–43. New York, NY: IEEE, 1995.

15. R. Eberhart, P. Simpson, and R. Dobbins. *Computational Intelligence PC Tools*. San Diego, CA: Academic Press Professional, Inc., 1996.

16. R. Lippmann. An introduction to computing with neural nets. *IEEE ASSP Mag.*, 4(2):4–22, 1987.

17. J. Hertz, A. Krogh, and G. R. Palmer. *Introduction to the Theory of Neural Computation*. Addison-Wesley, 1991.

18. D. D. Hawley, J. D. Johnson, and D. Raina. Artificial neural systems: A new tool for financial decision-making. *Fin. Anal. J.*, 46(6):63–2, 1990.

19. F. Rosenblatt. Principles of neurodynamics. 1962.

20. J. E. Dayhoff. *Neural Network Architectures: An Introduction*. New York: Van Nostrand Reinhold, 1990.

21. K. Mehrotra, C. K. Mohan, and S. Ranka. *Elements of Artificial Neural Networks*. Cambridge, MA: MIT Press, 1997.

22. C.-T. Lin and C. S. George Lee. Neural-network-based fuzzy logic control and decision system. *IEEE Trans. Comput.*, 40(12):1320–1336, 1991.

23. S. Horiawa, T. Furuhashi, S. Ouma, and Y. Uchikawa. A fuzzy controller using a neural network and its capability to learn expert's control rules. In *Proceedings of the International Conference on Fuzzy Logic and Neural Networks*. Japan, 103–106, 1990.

24. C. C. Lee. Intelligent control based on fuzzy logic and neural network theory. In *Proceedings of the International Conference on Fuzzy Logic and Neural Networks*. Iizuka, 759–764, 1990.

25. A. Traulsen, M. A. Nowak, and J. M. Pacheco. Stochastic payoff evaluation increases the temperature of selection. *J. Theor. Biol.*, 244:349–356, 2007.

26. X.-J. Chen, F. Fu, and L. Wang. Interaction stochasticity supports cooperation in spatial prisoner's dilemma. *Phys. Rev. E*, 78:051120, 2008.

27. F. Fu, C. Hauert, M. A. Nowak, and L. Wang. Reputation-based partner choice promotes cooperation in social networks. *Phys. Rev. E*, 78:026117, 2008.

28. M. A. Nowak and K. Sigmund. Evolution of indirect reciprocity by image scoring. *Nature*, 393:573–577, 1998.

29. J. Tanimoto: Simultaneously selecting appropriate partners for gaming and strategy adaptation to enhance network reciprocity in the prisoner's dilemma. *Phys. Rev. E*, 89(1):012106, 2014.

30. J. Li, C. Zhang, Q. Sun, Z. Chen, and J. Zhang, Changing the intensity of interaction based on individual behavior in the iterated prisoners dilemma game. *IEEE Trans. Evol. Comput.*, 21(4):506–517, 2017.

31. G. Szabó and C. Tőke. Evolutionary prisoner's dilemma game on a square lattice. *Phys. Rev. E*, 58:69–73, 1998.

32. L. A. Zadeh. Fuzzy sets. *Inform. Control*, 8(3):338–353, 1965.

33. C.-T. Lin and C. S. George Lee. Neural-network-based fuzzy logic control and decision system. *IEEE Trans. Comput.*, 40(12):1320–1336, 1991.

34. W. Xiaofan, L. Xiang, and C. Guanrong. *Network Science: An Introduction*. Beijing: Higher Education Press, 2012.

35. J. Zhang, C. Zhang, T. Chu, and M. Perc. Resolution of the stochastic strategy spatial prisoner's dilemma by means of particle swarm optimization. *PLoS One*, 6(7): e21787, 2011.

# Chapter 2

## The Spread of Multiple Strategies in the Complex Networks

Jianlei Zhang

## CONTENTS

Embodied in the evolutionary game theory, the strategy competition among agents has been studied in an effective way. In this chapter, we first present a collection of evolutionary game theoretic models that help to explore questions related to the strategy competition in nature. As an interdisciplinary topic, many subjects from modeling to analysis needs perfecting. For example, we still lack a deep understanding of the interplay between the individual heterogeneity (strategy, interactions, etc.) and the steady state of the system. However, the solutions for the above problem will make a crucial step toward the strategy competition problems in social systems. Here in this chapter, by the aid of establishing effective theoretical models to enrich the individual choices, some examples will be provided for a better understanding of these questions.

Specifically, we first provide some preliminaries about the swarm behavior and individual strategy in the gaming population (Sections 2.1–2.2). Then, we introduce the strategy competition among cooperation, defection, and speculation, hoping to shed light on how cooperation can be influenced by the introduction of speculation (Section 2.3). Next, how the speculation and loner behaviors influence the collective behaviors in multi-agent systems is introduced (Section 2.4). The approaches from mathematics, statistical physics, computer science, and engineering will be helpful in exploring the competing dynamics in the related populations involved in collective dilemma situations.

The main contribution of the chapter is in showing the role of the diversity of strategy choices of agents in the game playing. Individual heterogeneity is common in nature and social society of human. Whether and how this diversity affects the spreading of strategy, disease, information, opinion, and so on, is a fascinating topic. The definition and characteristics of some strategy choices, such as the insured cooperation, speculation, will be provided here. And their influences on the strategy spreading are analyzed generally and exactly, and the exact analytic solutions are also presented.

## 2.1    Evolutionary Game Theory

The origin and stability of cooperative or altruistic behaviors is a hot subject in social and behavioral sciences [1,2]. In multi-agent systems, some individuals will incur a cost to bring a benefit to another or the group. For example, in human activities like hunting for food, conserving common forestry or fisheries resources, altruistic behaviors are common and helpful for the collective benefits of the group. Hunting for food will bring benefits to the group, while it will make the hunter be at the peril of one's life in this activity. From the perspective of Darwinian evolution, it is not easy to understand this altruistic behavior because maximizing one's benefits stems from a survival instinct of individuals in nature. Here the altruistic agents are called as cooperators, while the selfish individuals are named as free riders or defectors here. In this sense, at least two individual choices (cooperation and defection) will exist in this type of collective actions. A cooperator will make a choice which (at least in the short term) brings the player with a smaller benefit than she would otherwise receive, but which helps the other player (do not cooperate) by increasing her payoff.

Thus, strategy competition is prevalent in many real-world situations involving cooperative dilemma, where defection will be the optimal strategy from the perspective of increasing the private interest, while mutual cooperation will be the best strategy for the group. Importantly, the strategy competition results will influence the overall performance of the group. Thus, investigating the influencing factors for the strategy competition, and even the control of strategy evolution will have wide range of applications in real social systems [3–5]. To framework the strategy competition and cooperative dilemma, the most prevailing framework is game theory and its extensions involving evolutionary context [6–10].

In the investigation of this topic, previous works have proposed many mechanisms to test their role in affecting the strategy competition and control the social dilemma in an efficient way. Among these works, a well-known model is that agents in real social systems are connected by complex interactions, not the well-mixed one in the ideal assumptions. In this sense, the strategy competition is dependant upon certain environmental conditions. One such condition that gets extensive attention and study is the use of a spatially structured population. The key concept of structured populations is: agents situate at the vertices of a network, which can be a regular lattice or a more complex structure. The edges denote links between players in terms of game dynamical interactions, in the framework of evolutionary game theory. Under this assumption, agents are limited to interact only with their neighboring partners to play evolutionary games. Vast studies and recent reviews can be found in References 11 and 12.

### 2.1.1   Typical game models

To characterize the interest conflicts between individuals and groups resulted by strategy competition, many theoretic models play a part in abstracting the essence of different social conflict scenarios. For example, the Prisoner's Dilemma game (PDG), Snowdrift game (SDG), Stag-Hunt game (SHG), Public Goods game (PGG), Rock–Paper–Scissors game (RPSG), Public Goods Game with Threshold (TPGG), and so on [13,14], are all often-used concrete game models. These specific models provide a framework to systematically analyze how the equilibrium state may be affected by the parameter changes. Since the following games will be employed in the majority of the works, we first provide an accurate description of them in what follows.

- *Prisoner's Dilemma Game*: The PDG is seen as a standard example of a game that shows why two completely rational agents may not cooperate, even if it appears that it is in their best interests to do so. As for its history, Albert W. Tucker formalized it with prison sentence rewards and named it as prisoner's dilemma [15]. Its details are presented as follows.

  The given scenario is that two members of a criminal gang are arrested and imprisoned by police. Each prisoner is in solitary confinement with no means of communicating with her partner in this criminal gang. The prosecutors lack sufficient evidence to convict the pair on the principal charge. In this situation, they hope to get both sentenced to a year in prison on a lesser charge. Meanwhile, the prosecutors offer each prisoner a bargain. Each prisoner is given the opportunity either to betray the other by testifying that the other committed the crime, or to cooperate with the other by remaining silent (Table 2.1). The offer is:

- If the criminal suspect $A$ and $B$ each betray the other, each of them serves two years in prison.

- If the criminal suspect $A$ betrays $B$ but $B$ remains silent, $A$ will be set free and $B$ will serve three years in prison (and vice versa).

- If the criminal suspects $A$ and $B$ both remain silent, both of them will only serve one year in prison.

For the sake of unification and without loss of generality, four parameters $R$, $T$, $P$, and $S$ are employed for describing the payoff matrix in theoretic studies in most works. Therefore, the payoff matrix of general symmetric two-player games is given by Table 2.2.

**TABLE 2.1:**     Prisoner's dilemma payoff matrix

|  | Prisoner B stays silent (cooperates) | Prisoner B betrays (defects) |
|---|---|---|
| Prisoner A stays silent (cooperates) | Each serves one year | Prisoner A: three years Prisoner B: goes free |
| Prisoner A betrays (defects) | Prisoner A: goes free Prisoner B: three years | Each serves two years |

**TABLE 2.2:**     Payoff matrix of general symmetric two-player games

|  | $C$ | $D$ |
|---|---|---|
| $C$ | $R$ | $S$ |
| $D$ | $T$ | $P$ |

**TABLE 2.3:** Payoff matrix of PDG

|     | *C*     | *D*  |
| --- | ------- | ---- |
| *C* | $b - c$ | $-c$ |
| *D* | $b$     | $0$  |

In the PDG (described by $T > R > P > S$) whose simplified payoff matrix is given by Table 2.3, where $b$ and $c$ ($b > c > 0$) indicate the benefits and costs of strategy cooperation, respectively. Standing on the position of self-centred players and deducing from the payoff matrix of PDG, the only stable evolutionary equilibrium by replicator dynamics equation is the domination state of defection.

- *Snowdrift Game*: The parameter settings of $T > R > S > P$ describes another type of collective dilemma and is named as SDG. As shown in Table 2.4, it is easy to see that the best choice for self-serving players is to do what contrasts with the opponent does. To gain larger benefits, the focal player needs to defect if the other player cooperates and to cooperate if the opponent chose defection. Here, the initial state including cooperators will drive the system to a stable coexistence equilibrium state at $x^* = \frac{P-S}{R-S-T+P}$, which indicate the fraction of cooperators.

- *Stag-Hunt Game*: Another order of $R > P > T > S$ (showed at Table 2.5) is often referred as SHG. For rational players with pursuit of larger benefits, the best choice is to do the same as what the opponent does. In this case, the final steady states depends on the initial strategy distribution among the participants.

- *Public Goods Game*: To expand the view of two-player games, it is natural to realize that many real-life dilemmas involve multiple players. Plenty of examples abound in many real-world situations involving cooperative behavior. For example, in situations referring to the common (forestry, fisheries, etc.) resources, game models involving multiple players will be more appropriate than two-player ones. In situations like these, each

**TABLE 2.4:** Payoff matrix of SDG

|     | *C*             | *D*     |
| --- | --------------- | ------- |
| *C* | $b - \frac{c}{2}$ | $b - c$ |
| *D* | $b$             | $0$     |

**TABLE 2.5:** Payoff matrix of SHG

|     | *C*     | *D*             |
| --- | ------- | --------------- |
| *C* | $b$     | $0$             |
| *D* | $b - c$ | $b - \frac{c}{2}$ |

group member will evenly enjoy the benefits distributed from the common goods, including those who pay no cost of providing the goods. This arouses the question of why agents participate in costly cooperative activities like warfare and risky hunting, now that they can gain the equally distributed common goods even without contributions.

Along this line, one of the frequently used multiple-agent-two-strategy models to describe the confusion of how cooperation arises is the PGG [16–18]. It focuses on the gains arising in multi-person interactive decision situations, when probably not all of the members choose to make contributions. Cooperation and defection are the two basic strategies that are usually at the heart of such social dilemma.

In a typical PGG played in an interacting system of size $N$, each player must independently and simultaneously make her decision to cooperate (contribute an amount $c$ to the public goods) or to defect (contribute nothing). The sum of the collected contributions is multiplied by a factor $r$ $(1 < r < N)$ and then redistributed to the $N$ players equally, irrespective of their individual contributions. For the group, the maximum benefit will be achieved if all players contribute maximally. Players are faced with the temptation of taking advantage of the common pool without contribution. In other words, any individual investment is a loss for the player because only a portion $r/N < 1$ will be repaid. Consequently, rational players invest nothing, hence to establish a social dilemma.

- *Threshold Public Goods Game*: Although the PGG is deemed as one of the often-used games in establishing the framework for strategy competition in multi-agent systems, there are still some social dilemmas for which a different game would be a more appropriate model. In many cases of a collective action, achieving of the group goal depends on the amount of common goods contributions. It is a common observation that many public goods contributed by collective actions are provided if contributions reach or exceed the required threshold of contributions; otherwise, no goods is provided [19,20]. Thus, a TPGG requires a minimum amount of contributions to be raised from a group of individuals for provision to occur [21–23].

  For instance, the cost of building a public dam (or road) or hunting in real social society cannot be provided by a single individual, since one agent has a finite capacity. The more individuals participating and cooperating, the less cost is required for each of them to finish the public project, given that the minimum threshold required to finish the project is achieved. In these cases, an effective project would require a minimum (threshold) number of contributions provided by the involved population for the project to be successfully accomplished. Researchers have examined the role of several factors, for example, incomplete information and identifiability of individual contributions, in influencing the public goods provision [24–26].

In a typical TPGG, each player in a group receives an endowment and individually decides how much of it to be contributed to a public goods system. If the group contribution exceeds a certain threshold, then the public goods is successfully provided by the group. And each player receives an equal reward, irrespective of her strategy. If the threshold is not reached, the game fails. Following an understandable assumption, contributions will not be returned to the players. Rational players face the temptation of selfishly free riding on others' contributions. Therefore, this rationale assumption of agents will lead to collective dilemmas and cause difficulties for the successful collection of the public goods.

To illustrate in a mathematical sense, it can be assumed that the TPGG is performed by a finite population of size $N$ ($N > 1$). Among the population, individuals are provided with identical endowment $c$, and each member independently decides how much (between all and none here) of her endowment to contribute to the common pool. After multiplying the accumulated contributions by an amplification coefficient $r$, each agent will gain an identical benefit distributed by the common goods, if the required threshold $T^*$ is reached. Note that $rc < T^* < rcN$ so that it is impossible for the threshold to be reached based solely upon the contribution of one player, but it is possible for it to be attained based upon the contributions of more than one player [27].

### 2.1.2   Strategy updating rules

Evolutionary game dynamics is closely related to how individuals update their strategies as time evolves. Many pervious works have widely employed imitation and replication in modeling the updating process [10,28,29].

The essence of replication rules is that a strategy with better performance has a higher replication rate. Two individuals, a focal individual and a referenced one, are sampled at random from the population. The focal player chooses a partner at random and decides whether to imitate her strategy with probability $p$ depending on a payoff comparison [30,31]. If both players have the same payoff, the focal individual chooses one of the two strategies randomly.

For example, the $p$ could be a linear function of the payoff difference [32], for example, $p = \frac{1}{2} + \omega_1 \frac{\pi_f - \pi_r}{\Delta \pi}$ or described by Fermi function [33]: $p = \frac{1}{1 + e^{\omega_2 (\pi_f - \pi_r)}}$. Here, $\pi_f$ and $\pi_r$, respectively, denote the payoffs of the focal individual and referenced one, and $\Delta \pi$ is the maximum payoff difference. $\omega_1$ and $\omega_2$ denote the noise or inverse temperature which control the selection intensity and take values in the range of $[0, 1]$ and $(0, \infty)$, respectively. The situation of $\omega_1 \to 0$ or $\omega_2 \to 0$ indicates all information is hidden by noise, yet the condition of $\omega_1 \to 1$ or $\omega_1 \to \infty$ denotes certain imitation rules.

For a system of infinite size, the deterministic replicator dynamics equation $\dot{x}_i = x_i (\pi_i - \langle \pi \rangle)$ has provided lots of insights for studying the strategy competition [34,35]. Here, $x_i$ is the fraction of strategy $i$ in the population, $\pi_i$ is the payoff or fitness of this strategy, and $\langle \pi \rangle$ is the average payoff in the whole population. If the payoff of strategy $i$ is below the average payoff, its density will decrease. If the payoff is above the mean payoff, then the corresponding density will increase. In general, $\pi_i$ depends on the strategy composition of population, that is to say, on the proportions of all other strategies $x_j$. Then, the average payoff $\langle \pi \rangle$ is the quadratic equation of the fraction $x_j$.

To establish an available model for the astoundingly strategy decision process of players is not easy, sparking heated debate about the related strategy updating rules is intriguing. Models for evolutionary games have traditionally assumed that players imitate their successful partners by the comparison of respective payoffs, raising the question of what happens if the game information is not easily available. Focusing on this yet-unsolved case, many works have been performed to establish a novel model for the updating states in a spatial population [36]. A key point is to steer by the needed payoffs in previous studies and focus on players' contact patterns. A new parameter of switching probability is thus proposed for determining the microscopic dynamics of strategy evolution. The gained results illuminate the conditions under which the steady coexistence of competing strategies is possible. Along this line, how other forms of switching probability promote coexistence is a topic that is worthy of further investigation.

### 2.1.3    Solutions for cooperative dilemma

Plenty of solutions have been put forward to explain the puzzle of the existing cooperative behaviors which damage ones' own benefits. For example, according to the kin selection theory, cooperation among individuals that are genetically related is possible. And, theory of direct reciprocity considers the possibility of the cooperative behaviors in bilateral long-term interactions [37,38]. Then, the theories of indirect reciprocity and signaling indicate how cooperation in large groups can emerge when cooperators can build a reputation [39]. Besides, punishment also plays a crucial role in the resolution of cooperative dilemma [40–43]. As already mentioned, the integration of the microscopic patterns of interactions among the individuals composing a large population into the evolutionary setting, broadens the study of cooperation in complex interacting systems.

Here is a very brief introduction about the complex network and the networked games:

- *Node*: the node is the principle unit of the network. A networks consists of a number of nodes connected by edges. In a typical setup of spatial evolutionary games, agents are assigned to the nodes of the network.

- *Neighbors*: two nodes are said to be neighbors if they are connected by a link or edge.

- *Link*: a link is a connection between two nodes in the networks. In the common setup of spatial evolutionary games, the edges denote links between the corresponding players in terms of game dynamical interactions.

- *Degree*: the degree of a node is the number of closest neighbors to which a node is interacted with. The average degree of the network is the mean of the individual degrees of all the nodes in the network.

- *Dynamics*: depending on the context, the word dynamics is used in the literature to refer to a temporal change of either the state or the topology of a network. In the common setup of spatial evolutionary games, it denotes the evolutionary game dynamics occurring on the interactions, being subject to the specific strategy updating rules or the introduced coevolution dynamics between networks and strategies.

A common framework is that each node in a graph carries one player, and edges determine who plays with whom [44–46]. Many studies are established in the framework of the typical spatial assumption. However, Zhang et al. [47] shifts their attention to an alternative new model of evolutionary PGGs over spatial groups. First, a set of players are distributed in spatial groups interacted with a network, and each individual plays the PGG with the other members in her group. Further, players hold certain expectations about the payoffs. Those who are dissatisfied with the current earnings have the chance to migrate to other linked groups. This setting is inspired by the phenomenon in real world that individuals may migrate for a larger benefit. The gained results are: larger average group size and milder degree of cooperative dilemma lead to lower cooperation level and larger average payoffs of the population. However, there are still some unsolved challenging theoretical problems, such as the stability of strategy competition among agents situating on multi-layer networks or with non-uniform migration scopes [48].

---

## 2.2 Potential Strategies

Cooperation and defection are the traditional and classical strategies provided for the players, irrespective of which game applies. Cooperators make contribution to the collective benefit at a personal cost or damage, while defectors make no contribution and selfishly take advantage of others' contributions.

Since individual heterogeneity is a common phenomenon in nature and society, and real agents always face multiple strategy choices in the competition with others involved in social dilemma situations. This is particulary true in the context of human cooperation where human decision-making is probably shaped by a wealth of individual factors. The two-strategy profile can not fully describe the individual heterogeneity.

Importantly, individual heterogeneity and social diversity are also well-known phenomena in nature and in social society of humans. It is a main focus whether and how the biodiversity affects the emergence and transmission of strategy, disease, information, opinion, and so on. The potential difficulties brought by individual heterogeneity in mathematical modeling, raise important challenges for existing theoretical models which have only considered simple individuals in games. By this consideration, more studies concerning with the individual heterogeneity or diversity and their possible coexistence, in the framework evolutionary game theory, are expected in the near future. By following this way, we gain more hints on cracking a series of perplexing puzzles about strategy competition in the real social society.

### 2.2.1   Tit for tat

Until now, there are many conceivable strategies for the repeated cooperative games. The best-known strategy for the iterated Prisoner's Dilemma (IPD), the Tit for Tat (TFT) strategy, requires that the player adopts cooperation in the first round and thereafter echoes what her opponent has played in the previous round. In plenty of simulations employing human subjects and with computer programs playing against each other, this TFT strategy has performed amazingly well, and has won both of Robert Axelrod's tournaments. The spirit of TFT strategy [49,50] is that one's actual move is equal to what the opponent did in the previous round.

### 2.2.2   Win-stay, lose-shift

According to another famous strategy called 'win-stay, lose-shift' (WSLS), the focal agent repeats the previous action if the resulting payoff has met her aspiration level. Otherwise, she changes her strategy. It was originally formulated and being described that responses to a situation that are followed by satisfaction will be strengthened but those that are followed by dissatisfaction will be weakened. Its rule is based on: retaining a successful option but to switch after a failure. In fact, it is a simple, convincing learning rule. For example, if animals feel ill after eating certain food they will avoid food of that taste in the future, if they find a food source empty they switch to another one, and so on.

The WSLS rule [51], as a general learning principle, can be applied to many types of repeated decision problems, such as the repeated Prisoner's Dilemma, a widely used paradigm for cooperation. Consider two players that are provided with two options: cooperation ($C$) and defection ($D$). Here, the initial action and a certain 'aspiration-level', that is, a minimum expectation concerning the payoff in each round, are the key points. If the expectation is satisfied the players retain their current strategy, otherwise, they change to the alternative one.

### 2.2.3   Always cooperate

The player always adopts the strategy of cooperation, irrespective of the actions of the opponents.

### 2.2.4   Always defect

The player always adopts the strategy of defection, irrespective of the actions of the opponents.

### 2.2.5   Out for tat

Many interesting studies with the the concept of conditional dissociation show that a strategy named as 'out for tat' (OFT) plays significant role in the coexistence of cooperation and defection. According to the definition, OFT means that an individual adopting strategy of cooperation will make response to the defection of opponents by merely leaving. It is clear that OFT will not tolerate the defection from opponents, but it will not be in retaliation for the opponents' strategy (apart from the action of taking out). This is different with the action from TFT [52]. Many works help to suggest that the option to leave a dissatisfactory partner in response to his behavior, can effectively promote cooperation in several settings. However, unsolved mysteries about the fundamental features that make this conditional dissociation take effect in this way still attracts the attention or researchers in this areas [53].

Segismundo et al. [54] provides some of the key conditions that can realize larger levels of cooperation among selfish groups based on the conditional dissociation. Importantly, an analytical formula is provided to estimate the expected degree of cooperation thus achieved. Specifically, the proposed model involves a system of agents who are paired to play an IPD. All individuals are endowed with the same ability to make reactions to the strategy previously chosen by the partner. And, without any other a priori constraint or exclusion, they may use any behavioral rule that is compatible with this capacity.

Notably, their results suggest that the strategy competition will finally evolve into either a non-cooperative or a partially cooperative regime, mainly in relation with the expected lifetime of players. Whenever the partially cooperative regime materializes, its long-run stability depends on the coexistence of defectors and 'OFT'-players. By the aid of extensive numerical simulations and analytical mean-field methods, results verify that conditional dissociation is needful for supporting cooperation, while other conditional strategies (such as TFT) remain present only in small population shares [54].

### 2.2.6   Voluntary participation

From another point of view, the two classical strategies of cooperation and defection indicate the obligatory participation of players. In many collective dilemmas, individuals have the option of voluntary participation in the games. Based on this idea, many different strategies (e.g., loner and punishment) have been proposed to investigate their potential roles in resolving the cooperative dilemma problems. Voluntary participation [55,56] allows players to adopt a risk-aversion strategy, also named as loner. For example, a loner may refuse to participate in an unpromising PGG but instead relies on a small but fixed payoff.

Christoph and Szabo [57] presents a simple but effective mechanism working under full anonymity, and shows that optional participation can foil exploiters and overcome the social dilemma. An interesting result is that in voluntary public goods interactions, cooperators and defectors have chance to coexist. This result can be found under diverse assumptions on interaction structure and adaptation mechanisms, leading usually to an endless cycle of the involved strategies. Since the domination of defection strategy is seldom here, voluntary participation thus effectively offers a way to resolve collective dilemmas.

### 2.2.7 Punishers

As mentioned, collective dilemmas are situations in which the best decision for an individual is not optimal, or is even harmful for the group. In collective actions faced by multi-agent systems, the role of moral emotions in the resolution of the collective dilemma game has recently become the matter of considerable debate. These emotions originate in social relationships and are built on reciprocal evaluations and judgments of the self and others. It is inspired by the fact that: many agents may exhibit social preferences, implying that they may not solely focus on self-interest but also care positively or negatively for the behaviors of the opponents. In theory and in experiments, punishment has turned out to be a simple and effective way to prevent free riding behaviors [58]. However, it is easy to think of scenarios in which punishment works will often be more complicated in natural settings. Nowadays, plenty of works have been established about whether and how various forms of punishment are effective in bringing about cooperation [59], peer punishment [42,60], pool punishment [61], and anti-social punishment [62].

Moreover, Balafoutas et al. [63] presents findings from the field experiment investigating whether punishment 'fits the crime' in real-life interactions. Their experiment was conducted by employing two largest train stations in Cologne, to ensure that interactions were most likely one-shot. Unlike in lab experiments, results suggest that altruistic punishment here does not increase with the severity of the violation, regardless of whether it is direct (confronting a violator) or indirect (withholding help). They also document growing concerns for counter-punishment as the severity of the violation increases, indicating that the marginal cost of direct punishment increases with the severity of violations. The evidence suggests that altruistic punishment may not provide appropriate incentives to deter large violations. Our findings thus offer a rationale for the emergence of formal institutions for promoting large-scale cooperation among strangers. Although many pioneering studies have explored the related concept of punishment [64], more theoretical work is needed to understand how the willingness to punish and retaliate punishment may have evolved and their implications for the evolution of strategies.

### 2.2.8   Speculators

Besides, other choices that may be adopted by agents in game playing, have also been proposed to enrich the potential strategy profiles for players, such as the insurance behavior against punishment [65].

This strategy sets are motivated from the widespread insurance mechanism in real social systems. First, experience in daily life shows that agents differ in personal features, such as economic status, demand for insurance in real world, and so on. When facing potential loss, agents may show heterogeneity in risk preferences. Some players tend to transfer their future loss to some insurance policy, which could provide some (part or full) compensation for their potential loss. Considering the uncertainty of completion state in some games, it is meaningful to provide insurance choice for the players to avoid or decrease the unfavorable loss in the game playing.

In the settings of Reference 65, the insurance against punishment is endowed with the defectors in public goods systems. A scenario of evolutionary competition between three competing strategies is devised in the promotion of public cooperation. Here, agents are confronted with ambiguous risks or losses, but meanwhile face the choice of being insured. In the TPGG, agents can buy an insurance that cover the cost of the potential loss. The model settings is established in an insurance deal, since the premium should not only be high enough to compensate the insurer for bearing the individual's risk. Meanwhile, it needs to be low enough so that a player would like to insure her risk for this premium.

Modeling the additional strategy options inspired by several real-life systems, has also evolved into a mushrooming avenue of research. Probing into more strategies is not the need for providing new ways of fostering cooperation in social dilemma, but also help the understanding about the coexistence of multiple competing strategies in nature. Proposing more competing strategies in the gaming populations is challenging and holds promises of exciting new discoveries.

---

## 2.3   The Competition among Cooperators, Defectors, and Insured Cooperators

Here, we introduce the work about the strategy competition among three strategies: cooperation, defection, and speculators [14]. Specifically, the two-strategy PGG is reframed as a cooperative dilemma among cooperators, insured cooperators, and free riders. The gap between theoretical model and economic behaviors in real social society can thus be filled by the mathematical analysis here to some extent.

## 2.3.1    Threshold public goods game

In a typical TPGG, each player in a group receives an endowment and independently decides her contribution to a common pool. If the resulted contribution exceeds some threshold, then the provision of public goods is successful. And each player receives an equal reward, with no relation with her performed strategy. Failing to reach the threshold will suggest the failure of the game, and the contributions will not be returned. Rational players intend to selfishly free ride on others' contributions, because contributors will benefit others at a cost to themselves.

In a finite population of size $N$ ($N > 1$), players are provided with the same endowment $c$. Each must independently decide how much (between all and none here) of her endowment to contribute. After multiplying the accumulated contributions by $r$, each player receives an identical benefit, if the required threshold $T^*$ is reached. The setting of $rc < T^* < rcN$ indicates that it is impossible for the threshold to be reached based solely upon the contribution from one player, but it is possible for it to be attained based upon the contributions of more than one player [27].

Facing with the potential loss, some cooperators prefer buying an insurance which can cover her possible loss. Here more accurately, they are renamed as insured cooperators. Other cooperators may disregard this insurance and readily bear the potential loss. Here, they are referred to as common cooperators. For the PGG played by $N$ players, the insured cooperators and common cooperators are both contributors and their numbers are denoted by $N_i$ (insured cooperators) and $N_c$ (common cooperators). Thus, the population includes $N_i + N_c$ contributors and $N_d$ defectors.

If the threshold is already achieved, two functions about payoffs are employed and described in the following two scenarios, respectively.

- *Scenario I*: If the resulted contribution exceeds the threshold, all the participants will share the benefit $T^*/N$ from the successful TPGG. Contributions above the threshold of provision will be wasted. Many examples conforming to this model can be seen, such as voting for building a public garden or dam. The residents are required to fill in a questionnaire, or vote, or petition the government to get the project approved. Whether the public project will be approved, depends on the amount of supporters and the required minimum numbers needed for successful action. Thus, the project gets approved only if enough voters are collected, and excess signatures play a meaningless role in affecting the results.

- *Scenario II*: If resulted contributions reach the threshold, contributions above the provision point will further result in collective benefit. Here the public goods can be assumed to be provided in an amount increasing with the contributions. For example, the distributed benefit can be in the linear form $rc(N_c + N_i)/N$, where more contributors will bring larger benefits. Returning to the earlier example, the residents hope to build the dam by voluntary contributions. The successful provision and observed efficiency of the project are positively related to the amount of contributions. More contributions exceeding the threshold helps to build a stronger dam.

To merge the two scenarios, a variable $\omega$ is introduced

$$U = \omega T^* + (1 - \omega)rc(N_c + N_i). \tag{2.1}$$

The variation of $\omega$ can realize the transition from scenario I (i.e., $\omega = 1$) to scenario II (i.e., $\omega = 0$). In between the two extremes, a mixed situation of these two scenarios will be seen.

## 2.3.2 The payoff calculation

For a player, the payoff depends upon her strategy and that of her opponents. Each player chooses to contribute all or nothing. Here the allocation rule is: each player reaps an equal benefit from the successful game, minus her cost related to her strategy. The contributors within a TPGG group include common cooperators (whose number is $N_c$) and insured cooperators (whose number is $N_i$).

When the threshold of common goods is attained,

$$rc(N_c + N_i) \geq T^*, \tag{2.2}$$

where $r$ denotes the amplification factor, and $T^*$ is the required threshold for the public goods provision to be realized.

For a group of size $N$ consisting of the three characters (i.e., cooperator $C$, defector $D$, and insured cooperator $I$), their payoffs are:

$$\begin{cases} P_c = \dfrac{U}{N} - c \\ P_d = \dfrac{U}{N} \\ P_i = \dfrac{U}{N} - c - \lambda \end{cases}. \tag{2.3}$$

The enhancement factor $r > 1$ means that if all cooperate, they are better off than if all defect. The required setting $r < N$ creates a situation that each individual is better off defecting than cooperating. The first term in the expression of payoffs is the benefit that the agent obtains from the public goods, while the second term denotes her cost by the adopted strategy. For a cooperator, the cost is the investment $c$ to the public goods. For an insured cooperator, the cost is the contribution $c$ to the common pool and her payment $\lambda$ to the insurance. Defectors make no contribution and exploit other players.

If the contributions can not reach the threshold,

$$rc(N_c + N_i) < T^*, \qquad (2.4)$$

the contributors lose their contributions and the common goods can not be provided finally.

Thus, their net payoffs are determined by

$$\begin{cases} P_c = -c \\ P_d = 0 \\ P_i = \varepsilon - c - \lambda \end{cases}. \qquad (2.5)$$

Compared with formula (2.3), each player is better off if the goods is provided than if it is not. For insured cooperators, they will be compensated by the insurance in this failing project. Thus, the payoff advantages of defectors over insured cooperators depend on the involved parameters: the cooperative contribution $c$, the compensation $\varepsilon$ ($\varepsilon > 0$) provided by the insurance, and the insurance cost $\lambda$.

For simplicity and without loss of generality, the cost $c$ from a contributor is set to 1. For $r > 0$, $rc(N_c + N_i) \geq T^*$ can be rewritten as $N - N_d \geq (T^*/r)$, and thus introduce $H = \text{ceil}[N - (T^*/r)]$. Notably, this ceiling function of $H$ returns the smallest integer greater than or equal to $N - (T^*/r)$. Substituting the function $H$ for $T^*$ thus yields a judgment: $N_d < H$ leads to the successful TPGG, and $N_d \geq H$ means the unsuccessful game. In the following study, the threshold value $H$ is set as the maximum number of defectors above which TPGG ends in failure. The resulting dynamics is closely related to the model parameters, as illustrated by the examples in Figure 2.1.

### 2.3.3   Evolutionary dynamics

Here a sufficiently large, well-mixed system of players are employed for game playing. From time to time, sample groups of $N$ players are chosen randomly and participate in a TPGG. Notably, the probability that two players in large populations ever encounter again can be neglected. The probability that there are $m$ defectors among the $N - 1$ other agents in the sample population of size $N$ in which a given player finds herself, is given by

$$\binom{N-1}{m} x_d^m (1 - x_d)^{N-1-m}. \qquad (2.6)$$

Here $x_d$ denotes the fraction of defectors in the population.

The expected payoff for a defector in such a group is

$$P_d = \sum_{m=0}^{H-2} \frac{\omega T^* + (1-\omega)r(N-1-m)}{N} \binom{N-1}{m} x_d^m (1-x_d)^{N-1-m}. \qquad (2.7)$$

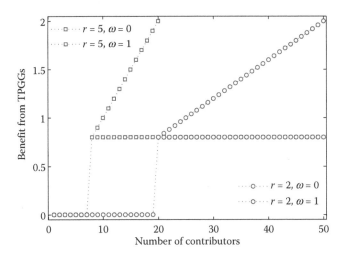

**FIGURE 2.1:** Diagrams illustrating four examples of TPGG, whose dynamics outcomes are closely related to the model parameter involved. Parameters here: $N = 50$, $T^* = 40$. The $x$-axis is indexed by the number of contributors (including cooperators and insured cooperators), and the $y$-axis represents the individual benefits from TPGG. Results show that, when $r = 5$, individuals can gain positive benefits from TPGG if there are at least 8 contributors. When $r = 2$, at least 20 contributors in one TPGG are needed to bring each participant with positive benefits. As mentioned, varying the parameter $\omega$ can transverse the model smoothly from scenario I (i.e., $\omega = 1$) to scenario II (i.e., $\omega = 0$) about the payoff functions in the TPGG after the threshold point has already been reached.

The payoff of a cooperator is given by

$$P_c = \sum_{m=0}^{H-1} \left[ \frac{\omega T^* + (1-\omega)r(N-m)}{N} - 1 \right] \binom{N-1}{m} x_d^m (1-x_d)^{N-1-m}$$

$$+ \sum_{m=H}^{N-1} (-1) \binom{N-1}{m} x_d^m (1-x_d)^{N-1-m}. \tag{2.8}$$

The payoff of an insured cooperator will thus be

$$P_i = \sum_{m=0}^{H-1} \left[ \frac{\omega T^* + (1-\omega)r(N-m)}{N} - 1 - \lambda \right] \binom{N-1}{m} x_d^m (1-x_d)^{N-1-m}$$

$$+ \sum_{m=H}^{N-1} (\varepsilon - 1 - \lambda) \binom{N-1}{m} x_d^m (1-x_d)^{N-1-m}. \tag{2.9}$$

Further, the advantage of one strategy over another is enslaved to the payoff difference between them.

*Competition between strategy C and I:*

**Theorem 2.1.** *There are two interior roots on the edge of ID when $\phi_1(x_{d,1}) + \varepsilon - \lambda - 1 > 0$ and $\varepsilon - (\lambda + 1) < 0$, one interior root on the edge of ID when $\phi_1(x_{d,1}) + \varepsilon - \lambda - 1 = 0$ and $\varepsilon - (\lambda + 1) < 0$ or when $\phi_1(x_{d,1}) + \varepsilon - \lambda - 1 > 0$ and $\varepsilon - (\lambda + 1) > 0$, and no interior root on the edge of ID when $\phi_1(x_{d,1}) + \varepsilon - \lambda - 1 < 0$, where $\phi_1(x_d) = \frac{[r(1-\omega) - N\varepsilon](N-H)}{N} \int_0^{1-x_d} t^{N-H-1}(1-t)^{H-1} dt + \frac{T^*}{N}\binom{N-1}{H-1}x_d^{H-1}(1-x_d)^{N-H}$ and $(x_{d,1}) = \frac{T^*(H-1)}{[r(1-\omega)-N\varepsilon](N-H)+T^*(N-1)}$.*

*Proof:*

$$P_c - P_i = \lambda - \varepsilon \sum_{m=H}^{N-1} \binom{N-1}{m} x_d^m (1 - x_d)^{N-1-m}. \tag{2.10}$$

*Then $\lim_{x_d \to 0} (P_c - P_i) \approx \lambda > 0$ and $\lim_{x_d \to 1} (P_c - P_i) \approx (\lambda - \varepsilon) < 0$.*

*Competition between strategy I and D:*

$$P_i - P_d = \varepsilon - \lambda - 1 + \sum_{m=0}^{H-1}\left[\frac{r(1-\omega)}{N} - \varepsilon\right]\binom{N-1}{m}x_d^m(1-x_d)^{N-1-m}$$
$$+ \frac{T^*}{N}\binom{N-1}{H-1}x_d^{H-1}(1-x_d)^{N-H}$$
$$= \varepsilon - \lambda - 1 + \frac{[r(1-\omega) - N\varepsilon](N-H)}{N}\int_0^{1-x_d} t^{N-H-1}(1-t)^{H-1} dt$$
$$+ \frac{T^*}{N}\binom{N-1}{H-1}x_d^{H-1}(1-x_d)^{N-H}. \tag{2.11}$$

*By introducing $\phi_1(x_d)$, Equation (2.11) can be rewritten as*

$$P_i - P_d = \varepsilon - \lambda - 1 + \phi_1(x_d), \tag{2.12}$$

*and hence,*

$$\frac{d\phi_1(x_d)}{dx_d} = \frac{[r(1-\omega) - N\varepsilon](N-H)}{N}\binom{N-1}{H-1}\left[-x_d^{H-1}(1-x_d)^{N-H-1}\right]$$
$$+ \frac{T^*}{N}\binom{N-1}{H-1}\left[(H-1)x_d^{H-2}(1-x_d)^{N-H}\right.$$
$$\left. - (N-H)x_d^{H-1}(1-x_d)^{N-H-1}\right] \tag{2.13}$$

*Provided that $0 < x_d < 1$ holds, the above Equation (2.13) keeps the same sign with $-[r(1-\omega) - N\varepsilon](N-H)x_d + T^*(H-1)(1-x_d) - T^*(N-H)x_d$.*

*Resolving the equation* $-[r(1-\omega)-N\varepsilon](N-H)x_d + T^*(H-1)(1-x_d) - T^*$
$(N-H)x_d = 0$ *yields*

$$x_{d,1} = \frac{T^*(H-1)}{[r(1-\omega)-N\varepsilon](N-H)+T^*(N-1)}. \tag{2.14}$$

*Consequently, both the maximum and minimum values of* $\phi_1(x_d)$ *exist, since* $\phi_1(x_d)$ *is continuous in* $[0,1]$. *Given that* $\frac{d\phi_1(x_d)}{dx_d} = 0$ *when* $x_d = x_{d,1}$, $\frac{d\phi_1(x_d)}{dx_d} > 0$ *if* $x_d < x_{d,1}$ *holds, and* $\frac{d\phi_1(x_d)}{dx_d} < 0$ *when* $x_d > x_{d,1}$, $P_i - P_d$ *reaches the maximum value at* $x_{d,1}$. *Then we can safely get* $x_{d,1} = \frac{T^*(H-1)}{T^*(N-1)-N\varepsilon(N-H)}$ *at* $\omega = 1$, $x_{d,1} = \frac{T^*(H-1)}{T^*(N-1)+(r-N\varepsilon)(N-H)}$ *at* $\omega = 0$.

*From Equation* (2.11), *we get* $\lim_{x_d \to 0}(P_i - P_d) \approx \frac{r(1-\omega)}{N} - \lambda - 1 < 0$ *and* $\lim_{x_d \to 1}(P_i - P_d) \approx \varepsilon - (\lambda + 1)$.

*To sum up, there are two interior roots on the edge of ID when* $\phi_1(x_{d,1}) + \varepsilon - \lambda - 1 > 0$ *and* $\varepsilon - (\lambda+1) < 0$, *one interior root on the edge of ID when* $\phi_1(x_{d,1}) + \varepsilon - \lambda - 1 = 0$ *and* $\varepsilon - (\lambda+1) < 0$ *or when* $\phi_1(x_{d,1}) + \varepsilon - \lambda - 1 > 0$ *and* $\varepsilon - (\lambda+1) > 0$, *and no interior root on the edge of ID when* $\phi_1(x_{d,1}) + \varepsilon - \lambda - 1 < 0$.    □

*Competition between strategy C and D:*

**Theorem 2.2.** *There are two interior roots on the edge of CD when* $\phi_2(x_{d,2}) > 1$, *one interior root when* $\phi_2(x_{d,2}) = 1$, *and no interior root when* $\phi_2(x_{d,2}) < 1$, *where* $\phi_2(x_d) = \frac{r(N-H)(1-\omega)}{N}\binom{N-1}{H-1}\int_0^{1-x_d} t^{N-H-1}(1-t)^{H-1}dt + \frac{T^*}{N}\binom{N-1}{H-1}x_d^{H-1}(1-x_d)^{N-H}$ *and* $x_{d,2} = \frac{T^*(H-1)}{r(1-\omega)(N-H)+T^*(N-1)}$.

*Proof: In analogy to the above methods, the sign of* $P_c - P_d$ *determines whether it pays to switch from defection to cooperation or vice versa, with* $P_c - P_d = 0$ *being the equilibrium condition.*

*Figure* 2.2 *illustrates three examples with respect to* $T^*$, *to help depicting the complicated situations of* $P_c - P_d$.

$$P_c - P_d = -1 + \sum_{m=0}^{H-1} \frac{r(1-\omega)}{N}\binom{N-1}{m}x_d^m(1-x_d)^{N-1-m}$$

$$+ \frac{T^*}{N}\binom{N-1}{H-1}x_d^{H-1}(1-x_d)^{N-H}. \tag{2.15}$$

*By employing*

$$\phi_2(x_d) = \frac{r(N-H)(1-\omega)}{N}\binom{N-1}{H-1}\int_0^{1-x_d} t^{N-H-1}(1-t)^{H-1}dt$$

$$+ \frac{T^*}{N}\binom{N-1}{H-1}x_d^{H-1}(1-x_d)^{N-H}, \tag{2.16}$$

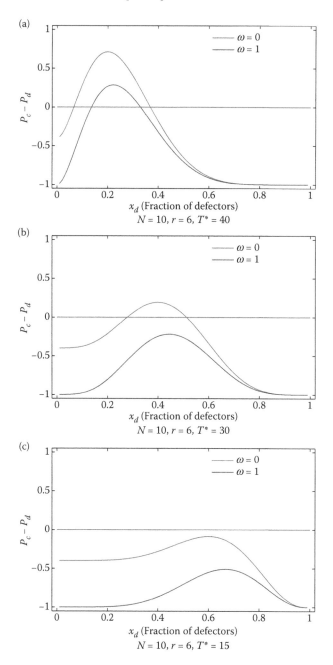

**FIGURE 2.2:** Examples illustrating the payoff difference $P_c - P_d$ between cooperators $P_c$ and defectors $P_d$, which is closely related to the required threshold $T^*$. The mentioned examples suggest that the possible roots of the $P_c - P_d$ will be: none, a unique or two roots situated in the interval $(0,1)$.

*Equation (2.15) can be reduced to*

$$P_c - P_d = -1 + \phi_2(x_d). \tag{2.17}$$

*Next,*

$$
\begin{aligned}
\frac{d\phi_2(x_d)}{dx_d} &= -\frac{r(1-\omega)(N-H)}{N} \binom{N-1}{H-1} \left[ -x_d^{H-1}(1-x_d)^{N-H-1} \right] \\
&\quad + \frac{T^*}{N} \binom{N-1}{H-1} \left[ (H-1)x_d^{H-2}(1-x_d)^{N-H} \right. \\
&\quad \left. - (N-H)x_d^{H-1}(1-x_d)^{N-H-1} \right] \\
&= \binom{N-1}{H-1} x_d^{H-2}(1-x_d)^{N-H-1} \left[ -\frac{r(1-\omega)(N-H)}{N} x_d \right. \\
&\quad \left. + \frac{T^*}{N}(H-1)(1-x_d) - \frac{T^*}{N}(N-H)x_d \right]
\end{aligned} \tag{2.18}
$$

$0 < x_d < 1$ *helps Equation (2.18) keep the same sign with* $-r(1-\omega)(N-H)x_d + T^*(H-1)(1-x_d) - T^*(N-H)x_d$. *Then,*

$$-r(1-\omega)(N-H)x_d + T^*(H-1)(1-x_d) - T^*(N-H)x_d = 0 \tag{2.19}$$

*gives rise to*

$$x_{d,2} = \frac{T^*(H-1)}{r(1-\omega)(N-H) + T^*(N-1)}. \tag{2.20}$$

$P_c - P_d = -1$ *when* $x_d = 0$, *and* $P_c - P_d = -1$ *when* $x_d = 1$. *Similarly,* $\phi_2(x_d)$ *is a continuous function in the interval of* $[0,1]$, *and thus both the maximum and minimum values of* $\phi_2(x_d)$ *can be found. Considering that* $\frac{d\phi_2(x_d)}{dx_d} > 0$ *if* $x_d < x_{d,2}$, *and* $\frac{d\phi_2(x_d)}{dx_d} < 0$ *if* $x_d > x_{d,2}$, $P_c - P_d$ *reaches its maximum value at* $x_{d,2}$. *In this case,* $\omega = 1$ *leads to* $x_{d,2} = \frac{H-1}{N-1}$, *and* $\omega = 0$ *results in* $x_{d,2} = \frac{T^*(H-1)}{r(N-H)+T^*(N-1)}$. *It thus follows that: there are two interior roots on the edge of CD when* $\phi_2(x_{d,2}) > 1$, *one interior root when* $\phi_2(x_{d,2}) = 1$, *and no interior root when* $\phi_2(x_{d,2}) < 1$.   □

In the continuous time model, the evolution of the fractions of the three strategies are given by

$$\dot{x}_k = x_k(P_k - \bar{P}), \tag{2.21}$$

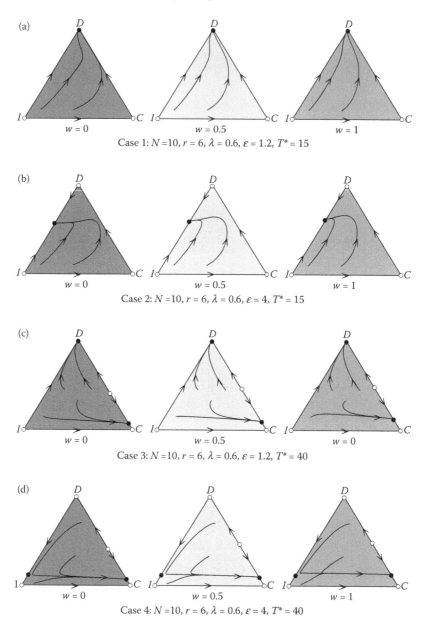

**FIGURE 2.3:** The dynamic results in different cases. The corners $C$ (cooperation), $D$ (defection), and $I$ (insured cooperation) are equilibrium pints. Open dots are unstable equilibrium points and closed dots are stable equilibrium points. In case 1, full $D$ is the only stable equilibrium while in the other three cases, other strategies may be the dominative ones.

where $k$ can be $c$, $d$, $i$, and $\bar{P} = x_c P_c + x_d P_d + x_i P_i$. Some possible cases of different parameters and the resulting game dynamics are demonstrated in Figure 2.3.

---

**Four Cases**

*Case 1:* $(\varepsilon - \lambda - 1 + \phi_1(x_{d,1}) < 0$, $\phi_2(x_{d,2}) - 1 < 0)$: Here, full defection equilibrium $(D)$ is the only stable and a global attractor. For an insured cooperator, her contribution and the cost for insurance cannot be totally reimbursed and thus she suffers negative payoffs when game fails. Thus, the dominant strategy equilibrium in case 1 is $D$.

*Case 2* $(\varepsilon - \lambda - 1 + \phi_1(x_{d,1}) > 0$, $\varepsilon - \lambda - 1 > 0$, and $\phi_2(x_{d,2}) - 1 < 0)$: Herein, there is a border equilibrium consisting of insured cooperation and defection. And this equilibrium is stable and a global attractor. Compared with case 1, the compensation $\varepsilon$ from insurance is increased and the resulted $\varepsilon < \lambda + 1$ will promote the survival of insured cooperators with higher payoffs than defectors.

*Case 3* $(\varepsilon - \lambda - 1 + \phi_1(x_{d,1}) < 0$, $\phi_2(x_{d,2}) - 1 > 0)$: Two border equilibrium points consisting of cooperation and defection exist. The one close to the full cooperation is a stable equilibrium and the other near full defection is unstable. Compared with case 1, increasing threshold $T^*$ leads to two stable equilibria here: full defection and the coexistence of cooperation and defection. Which equilibrium the system will evolve to depends on the initial states of the system.

*Case 4* $(\varepsilon - \lambda - 1 + \phi_1(x_{d,1}) > 0$, $\phi_2(x_{d,2}) - 1 > 0)$: Two stable border equilibria exist: one consisting of cooperation and defection, and the other consisting of insured cooperation and defection. Increasing $T^*$ leads to two equilibria on the edge of $CD$ here. Similar to case 3, lager $\omega$ will propel the equilibrium point on the edge of $CD$ to approach to the point of pure $D$.

---

*A Brief Summarization:* Summarizing the results above, the addition of insurance strategy encourages public contributions, interacting with the required threshold and the reimbursed compensation from insurance. Larger $T^*$ provides contributors with more payoff advantages than defectors. In addition, increasing the compensation $\varepsilon$ from insurance also dramatically alters the strategy competition outcomes. Insurance strategy can help contributors to receive higher payoffs than defectors. The insurance reduces or removes the risk of contributors' losing contributions when game fails. Sufficiently high compensation $\varepsilon$ will avoid the extinction of contributors, or even the possible dominance in the system. And, the allocation method of the public goods in successful games, also also notably influence the outcomes of strategy competition. Smaller $\omega$ enhances the payoff advantages of contributors over defectors. In this sense, the insurance strategy here suggests a positive role in unriddling the bewilderment of the 'Tragedy of the Commons'.

*Note*: The new strategy of insured cooperation can be applied in other specific collective dilemmas, besides the TPGG introduced here.

## 2.4     The Competition among Cooperators, Defectors, Speculators, and Loners

Voluntary participation allows individuals to adopt a risk-aversion strategy, termed loner. A loner refuses to participate in unpromising public enterprises and instead relies on a small but fixed payoff. How does the strategy compete when loners, speculators, cooperators, and defectors coexist in a system?

### 2.4.1     The introduction about loners

As mentioned, current research has also highlighted two factors boosting cooperation in public goods interactions: punishment on defectors [59,66] and the option to abstain from the PGG. Plenty of theoretical and empirical evidence points to the importance of punishment as a major factor for sustaining cooperation in PGGs [67]. As far as we know, voluntary participation [56] allows individuals to adopt a risk-aversion strategy, named as loner. A loner refuses to participate in unpromising PGG and instead relies on a small but fixed payoff.

In addition, the work introduced in section 2.2 proposes the choice of speculation, a kind of risk-aversion strategy [65]. Results indicate scenarios where speculation either leads to the reduction of the basin of attraction of the cooperative equilibrium or even the loss of stability of this equilibrium, if the costs of the insurance are lower than the expected fines faced by a defector.

Therefore, the joint roles of punishment, voluntary participation, and speculation in affecting the public goods provision is worthy studying [68]. Actually, agents often have multiple choices in decision-making according to the individual personality, especially when facing the potential punishment if acting as defectors. From the point of consciousness of risk prevention, agents probably perform different behaviors in real society. For example, resolute defectors will persist in their free riding behaviors, even taking the risk of being punished with a probability. Speculators would like to purchase an insurance covering the costs of punishment when caught defecting. However, timid loners are willing to conservatively and definitely gain a fixed payoff independent of the other players' decision. These choices can describe the possible attempts of agents involved in public goods actions in real-life situations. Thus, to study the strategy competition in biological scenarios, the heterogeneity of individual choices is an irrefutable fact and multiple choices deserve exploration. Here, based on the assumption that players can voluntarily decide whether to participate in the collective actions or not, the competition among four strategies are introduced.

The four strategies in the system are:

- *Cooperators* participate in the group and make contributions for the group;

- *Defectors* join in the game without contribution, at the risk of being caught with a certain probability. Here the specific establishment of punishment is neglected, and the two additional options (speculation and loner) are the focus;

- *Speculators* buy an insurance which can cover the costs of punishment when caught defecting. By paying a fixed cost for their insurance, speculators can defect without paying any fine for the punishment;

- *Loners* are reluctant to participate in the PGG, but would rather rely on a small but fixed payoff.

By means of a theoretical approach, the joint evolution of cooperation, defection, speculation, and loner will be investigated as follows, focusing on the question whether such model will allow the stable establishment of sizable levels of cooperation.

### 2.4.2   The model for collective dilemma

The investigation is based on the PGG, since it highlights the potential differences between individual interests and the social optimum. In the obligatory PGG model, the social dilemma can be considered as binary situations in which two strategies are available: either choose alternative $C$ (cooperation) in order to serve the public interest, or choose alternative $D$ (defection), which serves the immediate private interest.

A large population consisting of cooperators, defectors, speculators, and loners is employed. Specifically, each participant receives an equal benefit $rcx_c$, where $x_c$ is the fraction of cooperators ($x_c$) in the system. As usual, cooperators pay a fixed cost $c$ to the group. Defectors contribute nothing, and will be possibly caught and then confronted with punishment. The fine $\alpha$ reflects the product of the probability of being detected and the fine in cost of detective. Speculators pay a cost $\lambda$ for the insurance. Loners conservatively obtain a fixed payoff $\sigma$ without contribution.

### 2.4.3   Payoff calculation

A sufficiently large, well-mixed system of players is considered here. From time to time, sample groups of $N$ such players are chosen randomly and offered to join in a PGG. Here, $N_c$ denotes the number of cooperators and $N_l$ is the number of loners within such a group. The net payoffs of the four strategies are, respectively, given by

$$
\begin{cases}
P_c = \dfrac{rcN_c}{N - N_l} - c \\[2mm]
P_d = \dfrac{rcN_c}{N - N_l} - \alpha \\[2mm]
P_s = \dfrac{rcN_c}{N - N_l} - \lambda \\[2mm]
P_l = \sigma
\end{cases}
\tag{2.22}
$$

where $r$ is the amplification factor. Each contribution is multiplied by $r$ and the result is distributed among all participants (except loners) irrespective of their strategies. The first term in the expression represents the benefits, while the second term denotes cost. For cooperators, the cost is the investment $c$ to the public goods. For speculators, the cost is the payment $\lambda$ to the insurance.

To calculate the payoffs, it is necessary to derive the probability that $n$ of the $N$ sampled agents join in the game. When $n = 1$, the player has no other choice than to play as a loner and obtains payoff $\sigma$. This happens with the probability $x_l^{N-1}$, where $x_l$ is the fractions of loners. For a participator ($C$, $D$, or $S$), the probability of finding, among the $N-1$ other players in the sample, $n-1$ co-players joining the group ($n > 1$), is given by

$$
\binom{N-1}{n-1}(1 - x_l)^{n-1}(x_l)^{N-n}
\tag{2.23}
$$

The probability that $m$ of these players are cooperators is

$$
\binom{n-1}{m}\left(\frac{x_c}{x_c + x_d + x_s}\right)^m \left(\frac{x_d + x_s}{x_c + x_d + x_s}\right)^{n-1-m}
\tag{2.24}
$$

where $x_c$, $x_d$, $x_s$, respectively, denote the fractions of cooperators, defectors, and speculators in the population.

For simplicity and without loss of generality, the cost $c$ is set to 1. In the above case, the payoff for a defector is $rm/n - \alpha$, while the payoffs for a cooperator and a speculator are, respectively, specified by $r(m+1)/n - 1$ and $rm/n - \lambda$. Hence, the expected payoff for a defector in such a group is:

$$
\left(\frac{rm}{n} - \alpha\right)\sum_{m=0}^{n-1}\binom{n-1}{m}\left(\frac{x_c}{1 - x_l}\right)^m\left(1 - \frac{x_c}{1 - x_l}\right)^{n-m-1}
$$
$$
= \frac{r}{n}\cdot(n-1)\frac{x_c}{1 - x_l} - \alpha
$$

The payoff of a cooperator in a group of $n$ players is:

$$\left[\frac{r(m+1)}{n} - 1\right]\sum_{m=0}^{n-1}\binom{n-1}{m}\left(\frac{x_c}{1-x_l}\right)^m\left(1-\frac{x_c}{1-x_l}\right)^{n-m-1}$$

$$= \frac{r}{n}\cdot(n-1)\frac{x_c}{1-x_l} + \frac{r}{n} - 1$$

The payoff of a speculator in a group of $n$ players is:

$$\left(\frac{rm}{s} - \lambda\right)\sum_{m=0}^{N-1}\binom{n-1}{m}\left(\frac{x_c}{1-x_l}\right)^m\left(1-\frac{x_c}{1-x_l}\right)^{n-m-1}$$

$$= \frac{r}{n}\cdot(n-1)\frac{x_c}{1-x_l} - \lambda$$

The payoff of a loner is the fixed value of $\sigma$.
Then, the expected payoff for a defector in the population is

$$P_d = \sigma x_l^{N-1} + \sum_{n=2}^{N}\left[\frac{r}{n}\cdot(n-1)\frac{x_c}{1-x_l} - \alpha\right]\binom{N-1}{n-1}(1-x_l)^{n-1}(x_l)^{N-n}$$

$$= \sigma x_l^{N-1} + \frac{rx_c}{1-x_l}\left[1 - \frac{1-x_l^N}{N(1-x_l)}\right] - \alpha(1-x_l^{N-1}) \qquad (2.25)$$

In the continuous time model, the evolution of the four strategies proceeds according to

$$\dot{x}_i = x_i(P_i - \bar{P}), \qquad (2.26)$$

where $i$ can be $c$, $d$, $s$, $l$, and $\bar{P} = x_c P_c + x_d P_d + x_s P_s + x_l\sigma$.


## 2.4.4 Dynamics of strategy competition

For the replicator dynamics of three-strategy evolution, four scenarios are depicted in Figures 2.3–2.7. The advantage of one strategy over another depends on the payoff difference between them, hence

$$P_d - P_c = \sum_{n=2}^{N}\left[1 - \frac{r}{n} - \alpha\right]\binom{N-1}{n-1}(1-x_l)^{n-1}(x_l)^{N-n}$$

$$= 1 - \alpha + (r - 1 + \alpha)x_l^{N-1} - \frac{r}{N}\frac{1-x_l^N}{1-x_l}; \qquad (2.27)$$

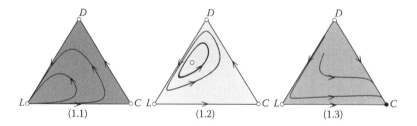

**FIGURE 2.4:** The evolution dynamics results of $T = (C, D, L)$ in the absence of speculation. (1.1): $r < 2 - 2\alpha$. (1.2): $r > 2 - 2\alpha$; and (1.3): $1 - r/N - \alpha < 0$. Parameters: $N = 5$, $\delta = 0.3$, and $r = 1.6$, $\alpha = 0.1$ for (1.1); $r = 3$, $\alpha = 0.1$ for (1.2); $r = 3$, $\alpha = 0.5$ for (1.3). Open dots are unstable equilibrium points and closed dots are stable equilibrium points. It suggests that three corners represent a rock–scissors–paper type heteroclinic cycle if $1 - r/N - \alpha > 0$ (cases 1.1 and 1.2) while pure cooperation is a global attractor if $1 - r/N - \alpha < 0$ (case 1.3).

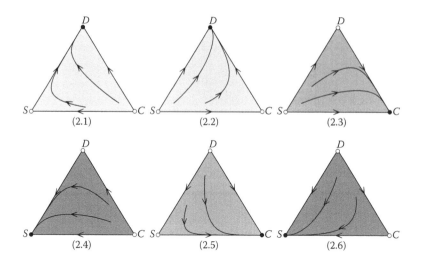

**FIGURE 2.5:** The evolution dynamics results of $T = (C, D, S)$ in the absence of defection. We consider six cases, which are discussed in cases 2.1 till 2.3 in the upper panel of Figure 2.5. Figure 2.5 focuses on the situation $\lambda - \alpha > 0$ implying that the fine for defectors is higher than the costs of cooperation. Lower panels of Figure 2.5 considers the opposite case $\lambda - \alpha < 0$, where defection is the dominating strategy. Results show that there is always a global attractor in the system, and the outcome of the game dynamics depends on model parameters. Parameters: $N = 5$, $r = 3$, $\delta = 0.3$, and $\alpha = 0.1$, $\lambda = 0.2$ for (2.1); $\alpha = 0.1$, $\lambda = 0.8$ for (2.2); $\alpha = 0.5$, $\lambda = 0.8$ for (2.3); $\alpha = 0.1$, $\lambda = 0.2$ for (2.4); $\alpha = 0.8$, $\lambda = 0.5$ for (2.5); $\alpha = 0.8$, $\lambda = 0.1$ for (2.6).

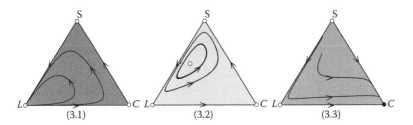

**FIGURE 2.6:** The evolution dynamics results of $T = (C, S, L)$ in the absence of speculation. (3.1): $r < 2 - 2\lambda$. (3.2): $r > 2 - 2\lambda$; and (3.3): $1 - r/N - \lambda < 0$. Parameters: $N = 5$, $\delta = 0.3$, and $r = 1.6$, $\lambda = 0.1$ for (3.1); $r = 3$, $\lambda = 0.1$ for (3.2); $r = 3, \lambda = 0.5$ for (3.3). It suggests that three corners represent a rock–scissors–paper type heteroclinic cycle if $1 - r/N - \lambda > 0$ (cases 3.1 and 3.2) while pure cooperation is a global attractor if $1 - r/N - \lambda < 0$ (case 3.3).

$$P_d - P_s = \sum_{n=2}^{N} [\lambda - \alpha] \binom{N-1}{n-1} (1 - x_l)^{n-1} (x_l)^{N-n}$$

$$= (\lambda - \alpha)(1 - x_l^{N-1}); \tag{2.28}$$

$$P_s - P_c = 1 - \lambda + (r - 1 + \lambda)x_l^{N-1} - \frac{r}{N} \frac{1 - x_l^N}{1 - x_l}. \tag{2.29}$$

In the above calculations, $N > 1$, $1 < r < N$ and $\alpha > 0$. The sign of $P_i - P_j$ determines whether it pays to switch from cooperation to defection or vice versa, $P_i - P_j = 0$ being the equilibrium condition, where $i, j$ can be strategy $C, D, S$, and $L$.

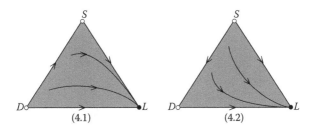

**FIGURE 2.7:** The evolution dynamics results of $T = (D, L, S)$ in the absence of cooperation (4.1), resulting game dynamics in the absence of speculation, where pure loners is the only global attractor in the system. Parameters: $N = 5$, $r = 3$, $\delta = 0.3$, and $\alpha = 0.4$, $\lambda = 0.1$ for (3); $\alpha = 0.4$, $\lambda = 0.1$ for (4.1); $\alpha = 0.1$, $\lambda = 0.4$ for (4.2).

When $\lambda \neq \alpha$, an interior point is that where four strategies coexist in the system. Given the following three assumptions, the conclusion is: at least one strategy will vanish in the system initialized from an interior point.

**Theorem 2.3.** *If $\lambda \neq \alpha$, at least one strategy will vanish in the system initialized from an interior point. Here, an interior point means the nonzero value of the fraction of each strategy.*

*Proof: the analysis results depend on different situations.*

1. When $\lambda \neq \alpha$, supposing $\lambda > \alpha$ (i.e., $P_d > P_s$), when $x_l \neq 0$. Supposing that there is a closed set, the subsequent evolving state of each initial state in this set also belongs to this set. So $x_c > 0$, $x_d > 0$, $x_s > 0$, and $x_l > 0$ in this closed set.

   (1.1) It is assumed that only one point $(x_c^*, x_d^*, x_s^*, x_l^*)$ in this closed set, satisfying $x_c^* > 0$, $x_d^* > 0$, $x_s^* > 0$, $x_c^* > 0$, and $\dot{x}_c^* = \dot{x}_d^* = \dot{x}_s^* = \dot{x}_l^* = 0$, thus

$$\begin{cases} \dot{x}_d^* = x_d^*(p_d^* - \bar{p}^*) \\ \dot{x}_s^* = x_s^*(p_s^* - \bar{p}^*) \end{cases} \tag{2.30}$$

   Here, $\dot{x}_d^* = \dot{x}_s^* = 0$ requires $\dot{p}_d^* = \bar{p}^* = \dot{p}_s^*$, which contradicts with $\dot{p}_d^* - \dot{p}_s^* > 0$. Thus, no interior stable point exists.

   (1.2) It is assumed that the interior domain is a limit cycle. In this case, the four types of players will get the same average payoffs, where $\bar{p}_c = \bar{p}_d = \bar{p}_s = \bar{p}_l$. However, $\bar{p}_d = \bar{p}_s$ contradicts with $p_d > p_s$, indicating that the closed set is not a limit cycle.

   (1.3) The following is to verify whether the interior domain is a chaos, where also $x_c > 0$, $x_d > 0$, $x_s > 0$, $x_l > 0$. By introducing the fraction of defections in a population consisting of defectors and speculators, $f = \dfrac{x_d}{x_d + x_s}$, thus

$$\dot{f} = \left(\frac{x_d}{x_d + x_s}\right)' = \frac{\dot{x}_d x_s - x_d \dot{x}_s}{(x_d + x_s)^2} = \frac{x_d x_s(p_d - p_s)}{(x_d + x_s)^2} > 0. \tag{2.31}$$

   Then, $\lim_{t \to \infty}\left(\dfrac{x_d}{x_d + x_s}\right) = 1$ and $x_s \to 0$.

   So, when $\lambda > \alpha$ there is no such a closed set, where the evolving state of each initial state which consist of these four strategies in this set also belongs to this set.

2. When $\lambda < \alpha$ and according to the results in (1), there is no internal domain.

3. When $\lambda = \alpha$ and thus $p_d = p_s$, the four-strategy system is reduced to the simplex $T = (C, D, L)$ or $T = (C, S, L)$, as discussed in the following.

A conclusion is that $\lambda = \alpha$ reduce the system to a three-strategy game, and $\lambda \neq \alpha$ will lead to the extinction of at least one strategy.

### 2.4.4.1 Scenario 1: The corners of the simplex $T = (C, D, L)$

**Theorem 2.4.** *If $r > 2 - 2\alpha$ holds, there there exists a threshold value of $x_1$ in the interval $(0, 1)$, above which $P_d - P_c < 0$.*

*Proof.* Here the function $G(x_l) = (1 - x_l)(P_d - P_c)$ is employed, which has the same roots as $P_d - P_c$. For $x_l \in (0, 1)$,

$$G(x_l) = (1 - x_l)(P_d - P_c)$$

$$= \left(1 - \frac{r}{N} - \alpha\right) - (1 - \alpha)x_l + (r - 1 + \alpha)x_l^{N-1} + \left(\frac{r}{N} + 1 - \alpha - r\right)x_l^N \tag{2.32}$$

$$G'(x_l) = (\alpha - 1) + (N - 1)(r - 1 + \alpha)x_l^{N-2} + N\left(\frac{r}{N} + 1 - \alpha - r\right)x_l^{N-1} \tag{2.33}$$

Note that $G(1) = G'(1) = 0$,

$$G''(1) = (N - 1)(N - 2)(r - 1 + \alpha)x_l^{N-3} + N(N - 1)\left(\frac{r}{N} + 1 - \alpha - r\right)x_l^{N-2} \tag{2.34}$$

$$G''(1) = (N - 1)(2 - 2\alpha - r) \tag{2.35}$$

We have

$$G(x_l) \simeq G(1) + G'(1)(z - 1) + \frac{1}{2}G''(1)(z - 1)^2$$

$$= \frac{1}{2}(N - 1)(2 - 2\alpha - r)(1 - x_l)^2. \tag{2.36}$$

For $r > 2 - 2\alpha$, $\lim_{x_l \to 1^-} G(x_l) < 0$,

$$G''(x_l) = x_l^{N-3}(N - 1)[(N - 2)(r - 1 - \alpha) + x_l(r + N - N\alpha - Nr). \tag{2.37}$$

Since $G''(x_l)$ changes sign at most once in the interval $(0, 1)$, there exists a threshold value of $x_l$ in the interval $(0, 1)$, above which $P_d - P_c < 0$. So,

$$\begin{cases} G(x_l) = (1 - x_l)(P_d - P_c) \\ G(0) = 1 - \dfrac{r}{N} - \alpha \\ G(1) = 0 \end{cases} \tag{2.38}$$

As illustrated in Figure 2.4, the game dynamics takes on three qualitatively different cases.

*Case 1.1*　$(1 - r/N - \alpha > 0, \text{ i.e., } G(0) > 0)$:

$$\lim_{x_l \to 1^-} G(x_l) = \frac{1}{2}(N - 1)(2 - 2\alpha - r)(1 - x_l)^2. \qquad (2.39)$$

*When $r < 2 - 2\alpha$, $G(x_l) > 0$, $x_l \in (0, 1)$, the three corners represent a rock–scissors–paper type heteroclinic cycle, and there is no stable equilibrium of the game dynamics in this case.*

*Case 1.2*　$(1 - r/N - \alpha > 0, r > 2 - 2\alpha, G(1^-) > 0)$: *the three corners represent a heteroclinic cycle. It is a center surrounded by closed orbits. Similar to case 1.1, there is no stable equilibrium of the game dynamics in this case.*

*Case 1.3*　$(1 - r/N - \alpha < 0, \text{i.e., } r > 2 - 2\alpha)$: *In this case, for all $x_s$, pure speculation (S) and pure defection (D) are both unstable equilibria of the game dynamics. The cooperation equilibrium (C) is stable and in fact a global attractor.*

*Summarizing the three cases in this scenario corresponding to the simplex $T = (C, D, L)$, the three corners represent a rock–scissors–paper type heteroclinic cycle if $1 - r/N - \alpha > 0$ (cases 1.1 and 1.2) while pure cooperation is a global attractor if $1 - r/N - \alpha < 0$ (case 1.3).*　□

**Proposition 2.5.** *When $T = (C, D, L)$, under the replicator dynamics of (2.26), it holds that*

*if $1 - r/N - \alpha > 0$ and $r < 2 - 2\alpha$, there is no inner fixed point in $T$;*

*if $1 - r/N - \alpha > 0$ and $r > 2 - 2\alpha$, there is one inner fixed point in $T$;*

*if $1 - r/N - \alpha < 0$, full $C$ is only stable fixed point in $T$.*

*Proof: When $r > 2 - 2\alpha$, there exists a fixed point $x_l \in (0, 1)$ that $P_d = P_c$. From Equation (2.25), the only $x_c$ and $x_d = 1 - x_l - x_c$ can be gained, hence there is one inner fixed point in $T$. If $1 - r/N - \alpha > 0$ and $r < 2 - 2\alpha$, $P_d > P_c$ for all $x_l \in (0, 1)$, so there is no fixed point in $T$. If $1 - r/N - \alpha < 0$, we have $r > 2 - 2\alpha$, $(N > 2)$. Then it must be true that $P_c > P_d$, so full $C$ is only stable fixed point in $T$.*　□

#### 2.4.4.2　Scenario 2: The corners of the simplex $T = (C, D, S)$

$$\begin{cases} P_d - P_c = 1 - \alpha - \dfrac{r}{N} \\ P_d - P_s = \lambda - \alpha \\ P_c - P_s = \lambda + \dfrac{r}{N} - 1 \end{cases} \qquad (2.40)$$

*Case 2.1* ($\lambda - \alpha > 0$, $1 - \alpha - r/N > 0$ and $1 - \lambda - r/N > 0$): Here, pure cooperation and pure speculation are both unstable equilibria of the game dynamics. Full defection equilibrium ($D$) is a stable and global attractor.

*Case 2.2* ($\lambda - \alpha > 0$, $1 - \alpha - r/N > 0$ and $1 - \lambda - r/N < 0$): Pure cooperation and pure speculation are both unstable equilibria of the system. Pure defection ($D$) is stable and a global attractor. The difference between case 2.1 and case 2.2 is that for a system only contains cooperators and speculators, pure cooperation is the attractor in case 2.2 while pure speculation is the attractor in case 2.1.

*Case 2.3* ($\lambda - \alpha > 0$, $1 - \alpha - r/N < 0$, and $1 - \lambda - r/N < 0$): Pure defection and pure speculation are both unstable equilibria of the gaming system. Pure cooperation is a stable and global attractor.

*Case 2.4* ($\lambda - \alpha < 0$, $1 - \alpha - r/N > 0$, and $1 - \lambda - r/N > 0$): Pure speculation is the only stable and global attractor.

*Case 2.5* ($\lambda - \alpha < 0$, $1 - \alpha - r/N < 0$, and $1 - \lambda - r/N < 0$): Pure cooperation is the only stable and global attractor.

*Case 2.6* ($\lambda - \alpha < 0$, $1 - \alpha - r/N < 0$, and $1 - \lambda - r/N > 0$): Pure speculation is the only stable and global attractor. The difference between case 2.6 and 2.4 is that when the population only contains cooperators and defectors, pure cooperation is the attractor in case 2.6 while pure defection is the attractor in case 2.4.

Summarizing the six cases in scenario 2 corresponding to the simplex $T = (C, D, S)$, there is always a global attractor in the system.

**Proposition 2.6.** *When $T = (C, D, S)$, under the replicator dynamics of (6.5), it holds that*

*if $\lambda - \alpha > 0$ and $1 - \alpha - r/N > 0$: full D is only stable fixed point in T;*

*if $1 - \alpha - r/N < 0$ and $1 - \lambda - r/N < 0$: full C is only stable fixed point in T;*

*if $\lambda - \alpha < 0$ and $1 - \lambda - r/N >$: full S is only stable fixed point in T;*

*Proof: When $x_l = 0$, if $1 - \alpha - r/N > 0$, $P_d > P_c$; if $\lambda - \alpha > 0$, $P_d > P_s$, therefore, if $x_d > 0$, $P_d > \bar{P}$. That means full D ($x_d = 1$) is only stable fixed point in T.*
    *When $x_l = 0$, if $1 - \alpha - r/N <$, $P_c > P_d$; if $1 - \lambda - r/N < 0$, $P_c > P_s$, therefore, if $x_c > 0$, $P_c > \bar{P}$. That means full C ($x_c = 1$) is only stable fixed point in T.*
    *When $x_l = 0$, if $\lambda - \alpha < 0$, $P_s > P_d$; if $1 - \lambda - r/N > 0$, $P_s > P_c$, therefore, if $x_s > 0$, $P_s > \bar{P}$. That means full S ($x_s = 1$) is only stable fixed point in T.* $\quad\square$

### 2.4.4.3    Scenario 3: The corners of the simplex $T = (C, L, S)$

It is easily observed that $x_l = 0$ leads to $P_c - P_s = \lambda - 1 < 0$. Thus, the three corners represent a rock–scissors–paper type heteroclinic cycle. There is no stable equilibrium in this case.

**Proposition 2.7.** *When $T = (C, S, L)$, under the replicator dynamics of (2.26), it holds that*

*if $1 - r/N - \lambda > 0$ and $r < 2 - 2\lambda$, there is no inner fixed point in $T$; if $1 - r/N - \lambda > 0$ and $r > 2 - 2\lambda$, there is one inner fixed point in $T$; if $1 - r/N - \lambda < 0$, full $C$ is only stable fixed point in $T$.*

*Proof:* Replacing $\alpha$ with $\lambda$ leads similar results with proposition 5.          □

### 2.4.4.4    Scenario 4: The corners of the simplex $T = (D, L, S)$

> *Case 4.1 $(\lambda - \alpha < 0)$:* In this case, pure loners is the only stable and in fact the only global attractor.
>
> *Case 4.2 $(\lambda - \alpha < 0)$:* Still, pure loners remains the only stable and in fact the only global attractor. The difference between case 4.1 and 4.2 is that: for a system only contains speculators and defectors, pure speculation is the attractor in case 4.1 while pure defection is the attractor in case 4.2.

Summarizing the two cases in scenario 4 corresponding to the simplex $T = (C, D, S)$, pure loner is the only global attractor in the system.

**Proposition 2.8.** *When $T = (S, D, L)$, under the replicator dynamics of (2.26), it holds that full $L$ is only stable fixed point in $T$;*

*Proof:* When $x_c = 0$, $P_l - P_d = (\alpha + \sigma)(1 - N_l^{N-1}) > 0$ and $P_l - P_s = (\lambda + \sigma)(1 - N_l^{N-1}) > 0$, full $L$ $(x_l = 1)$ is only stable fixed point in $T$.          □

*A brief summarization:* The extensive research here mainly focuses on the diversity of strategies. In the obligatory public goods model, the often-provided strategies are defection and cooperation. Daily experiences inspire us that when facing potential punishment as a defector, speculation, and optional participation are potential choices by some individuals. This is also a characteristic of individual diversity in a wide range of real-world situations.

Here the dynamic analysis of the gaming system is based on replicator dynamics for four strategies: $C$ (cooperators), $D$ (defectors), $S$ (speculators), and $L$ (nonparticipants) is established. The work here provides clear conditions under which speculation either leads to the reduction of the basin of attraction of the cooperative equilibrium or even the loss of stability of this equilibrium.

Specifically, the evolutionary fate of the system depends on special assumptions about model parameters. When starting from the three-strategy state, the observed domination of some strategy or a rock–paper–scissors type of cycle suggests that the additional strategy can significantly alter the strategy competition results. Here, the option to abstain from the game leads to the decline of free riders and the spreading of cooperators. Further, cooperation can also be promoted to be an equilibrium by moderate values of punishment $\alpha$ and cost of insurance $\lambda$ in the absence of loner (scenarios 2). Besides, cooperation has few chance to dominate the system in the competition with speculation and loner strategy, even though defection is absent (scenarios 3). When the initial state consists of the four strategies, at least one strategy fails to survive at the steady state.

## 2.5   Conclusion

Individual choices take heterogeneity and diversity in real social society. As mentioned, different scenarios of cooperative dilemma can be suitably framed by specific game models, such as the TPGG in section 2.3–2.4. When being involved in the specific dilemmas, individual may face different choices. For example, when punishment exists in the system, the choice of being insured is meaningful for the players who aim to free ride. Along this line, the following may be the potential problems in the mechanism of punishment.

a. Who will carry out the punishment?

b. Who will bear the cost for punishing others?

c. Under what conditions the punishment takes effect?

d. Who will buy the insurance?

e. Does retaliation occur when defectors are punished?

Summarizing, though cooperation and defection are typical strategies in the evolutionary game theory, the real social system based on intelligent agents often face the diversity of individual choices. Only modeling and incorporating them into the two-strategy profile, we can approach the real situations about strategy competition. Nowadays, many experimental study by employing real people as the subjects, have been performed and the results are more convictive. However, how to model the characteristics of real people in decision-making is not easy, and especially the following theoretical analysis based on them. To fill in the gap between experimental study and mathematical modeling, more attention is expected to put on this subject.

Besides, an interesting future direction would be to address whether the presence of more strategy options altogether affect the dynamics of behaviors in the field of collective cooperation.

## 2.6   Glossary

**Players:** The agents who participate in the game.
**Strategy:** The choice adopted by agents who participate in a game.
**Payoffs:** The benefits an agent gains after she plays a game with her partners.
**Strategy Competition:** Players will update their strategies for higher payoffs, leading to the spreading competition of the provided strategies.
**Steady State:** The strategy distribution of the multi-agent system will reach fixed after the evolution driven by strategy switching.

## References

1. A. Robert. *The Evolution of Cooperation*. Basic Books, New York, 1984.

2. R. Axelrod and D. Dion. The further evolution of cooperation. *Science*, 242:1385–1390, 1988.

3. W. D. Hamilton. The evolution of altruistic behavior. *Am. Nat.*, 97:354–356, 1963.

4. A. N. Martin, B. Sebastian, and M. M. Robert. Robustness of cooperation. *Nature*, 379:125–126, 1996.

5. K. Sigmund and M. A. Nowak. Evolutionary game theory. *Curr. Biol.*, 9:R503–R505, 1999.

6. E. Fehr and U. Fischbacher. The nature of human altruism. *Nature*, 425:785–791, 2003.

7. K. Sigmund, H. De Silva, A. Traulsen, and C. Hauert. Social learning promotes institutions for governing the commons. *Nature*, 466:861–863, 2010.

8. L. Conradt. When it pays to share decisions. *Nature*, 471:40–41, 2011.

9. A. N. Martin and S. Karl. Bacterial game dynamics. *Nature*, 418:138–139, 2002.

10. J. Zhang, X. Chen, C. Zhang, L. Wang, and T. Chu. Elimination mechanism promotes cooperation in coevolutionary prisoner's dilemma games. *Phys. A*, 389:4081–4086, 2010.

11. G. Szabó and G. Fáth. Evolutionary games on graphs. *Phys. Rep.*, 446:97–216, 2007.

12. H. Ohtsuki, C. Hauert, E. Lieberman, and M. A. Nowak. A simple rule for the evolution of cooperation on graphs and social networks. *Nature*, 441:502–505, 2006.

13. G. Cimini and A. Sánchez. Learning dynamics explains human behaviour in prisoner's dilemma on networks. *J. R. Soc. Interface*, 11(94):20131186, 2014.

14. C. Zhang, J. Zhang, and M. Cao. How insurance affects altruistic provision in threshold public goods games. *Sci. Rep.*, 5, 2015, srep09098. https://www.nature.com/articles/srep09098.

15. W. Poundstone. *Prisoner's Dilemma*. Doubleday, New York, 1992.

16. H. Brandt, C. Hauert, and K. Sigmund. Punishing and abstaining for public goods. *Proc. Natl Acad. Sci. USA*, 103:495–497, 2006.

17. D. Semmann, H. Krambeck, and M. Milinski. Volunteering leads to rock-paper-scissors dynamics in a public goods game. *Nature*, 425:390–393, 2003.

18. F. C. Santos, M. D. Santos, and J. M. Pacheco, Social diversity promotes the emergence of cooperation in public goods games, *Nature*, 454:213–216, 2008.

19. M. Milinski, R. D. Sommerfeld, H.-J. Krambeck, F. A. Reed, and J. Marotzke. The collective-risk social dilemma and the prevention of simulated dangerous climate change. *Proc. Natl Acad. Sci.*, 105(7):2291–2294, 2008.

20. J. Wang, F. Fu, T. Wu, and L. Wang. Emergence of social cooperation in threshold public goods games with collective risk. *Phys. Rev. E*, 80(1):016101, 2009.

21. C. B. Cadsby, Y. Hamaguchi, T. Kawagoe, E. Maynes, and F. Song. Cross-national gender differences in behavior in a threshold public goods game: Japan versus Canada. *J. Econ. Psychol.*, 28:242–260, 2007.

22. B. Cadsby Charles, C. Rachel, M. Melanie, and E. Maynes. Step return versus net reward in the voluntary provision of a threshold public good, An adversarial collaboration. *Public Choice*, 135:277–289.

23. T. A. C. Rachel, and M. B. Marks. Step returns in threshold public goods A meta-and experimental analysis. *Exp. Econ.*, 2(3):239–259, 2000.

24. M. Archetti and I. Scheuring. Coexistence of cooperation and defection in public goods games. *Evolution*, 65(4):1140–1148, 2011.

25. F. C. Santos and J. M. Pacheco. Risk of collective failure provides an escape from the tragedy of the commons. *Proc. Natl Acad. Sci.*, 108:(26):10421–10425, 2011.

26. G. Boza and S. Számadó. Beneficial laggards: Multilevel selection, cooperative polymorphism and division of labour in threshold public good games. *BMC Evol. Biol.*, 10:(1):336, 2010.

27. B. C. Charles and E. Maynes. Voluntary provision of threshold public goods with continuous contributions experimental evidence. *J. Public Econ.*, 71(1):53–73, 1999.

28. K. H. Schlag. Which one should i imitate? *J. Math. Econ.*, 31(4):493–522, 1999.

29. M. V. Pashkam, M. Naber, and K. Nakayama. Unintended imitation affects success in a competitive game. *Proc. Natl Acad. Sci. USA*, 110(50):20046–20050, 2013.

30. T. Galla. Imitation, internal absorption and the reversal of local drift in stochastic evolutionary games. *J. Theor. Biol.*, 269:46–56, 2011.

31. A. Sánchez, D. Vilone, J. Ramasco, and M. San Miguel. Social imitation versus strategic choice, or consensus versus cooperation, in the networked prisoner's dilemma. *Phys. Rev. E*, 90(2):022810, 2014.

32. A. Traulsen, J. C. Claussen, and C. Hauert: Coevolutionary dynamics: From finite to infinite populations. *Phys. Rev. Lett.*, 95(238701), 2005.

33. L. E. Blume. The statistical mechanics of best-response strategy revision. *Games Econ. Behav.*, 5(387), 111–145, 1995.

34. H. Abbass, G. Greenwood, and E. Petraki. The n-player trust game and its replicator dynamics. *IEEE Trans. Evol. Comput.*, 20(3):470–474, 2016.

35. I. M. Bomze. Lotka-volterra equation and replicator dynamics: new issues in classification. *Biol. Cybern.*, 72:447–453, 1995.

36. M. Cao. J. Zhang, C. Zhang, and F.J. Weissing. Crucial role of strategy updating for coexistence of strategies in interaction networks. *Phys. Rev. E*, 91(4):042101, 2015.

37. M. A. Nowak, Five rules for the evolution of cooperation, *Science*, 314:1560–1563, 2006.

38. H. Ohtsuki and M. A. Nowak. Direct reciprocity on graphs, *J. Theor. Biol.*, 247:462–470, 2007.

39. H. Brandt and K. Sigmund. Indirect reciprocity,image scoring, and moral hazard. *Proc. Natl Acad. Sci. USA*, 102:2666–2570, 2005.

40. T. Clutton-Brock and G. A. Parker. Punishment in animal societies. *Nature*, 373:209–216, 1995.

41. O. Gurerk, B. Irlenbusch, and B. Rockenbach. The competitive advantage of sanctioning institutions. *Science*, 312:108–111, 2006.

42. C. Hauert, A. Traulsen, H. Brandt, M. A. Nowak, and K. Sigmund. Via freedom to coercion: The emergence of costly punishment. *Science*, 316:1905–1907, 2007.

43. S. Gächter, E. Renner, and M. Sefton. The long-run benefits of punishment. *Science*, 322:1510, 2008.

44. A.-L. Barabási and R. Albert. Emergence of scaling in random networks. *Science*, 286:509–512, 1999.

45. C.Y. Zhang, J.L. Zhang, G.M. Xie, and L. Wang. Coevolving agent strategies and network topology for the public goods games. *Eur. Phys. J. B*, 80:217–222, 2011.

46. C. Zhang, J. Zhang, G. Xie, and L. Wang. Diversity of game strategies promotes the evolution of cooperation in public goods games. *Europhys. Lett.*, 90:68005, 2010.

47. J. Zhang, C. Zhang, and T. Chu. The evolution of cooperation in spatial groups. *Chaos Solit. Fract.*, 44:131–136, 2011.

48. C. Zhang, J. Zhang, and G. Xie. Evolution of cooperation among game players with non-uniform migration scopes. *Chaos Solit. Fract.*, 59:103–111, 2014.

49. M. A. Nowak and K. Sigmund. Tit for tat in heterogeneous population. *Nature*, 355:250–253, 1992.

50. S. K. Baek and B. J. Kim. Intelligent tit-for-tat in the iterated prisoner's dilemma game. *Phys. Rev. E*, 78:011125, 2008.

51. A. N. Martin and S. Karl. A strategy of win-stay, lose-shift that outperforms tit-for-tat in the prisoner's dilemma game. *Nature*, 364:56–58, 1993.

52. N. Hayashi. From tit-for-tat to out-for-tat. *Sociol. Theory Methods*, 8:19–32, 1993.

53. E. Hauk. Multiple prisoner's dilemma games with (out) an outside option an experimental study. *Theory Decis.*, 54(3):207–229, 2003.

54. S. I. Segismundo, R. Izquierdo Luis, and F. Vega-Redondo. The option to leave, conditional dissociation in the evolution of cooperation. *J. Theor. Biol.*, 267(1), 76–84, 2010.

55. C. Hauert, S. D. Monte, J. Hofbauer, and K. Sigmund. Volunteering as Red Queen mechanism for cooperation in public goods game. *Science*, 296:1129–1132, 2002.

56. C. Hauert, S. D. Monte, J. Hofbauer, and K. Sigmund. Replicator dynamics in optional public goods games. *J. Theor. Biol*, 218:187–194, 2002.

57. H. Christoph and G. Szabo. Prisoner's dilemma and public goods games in different geometries compulsory versus voluntary interactions, *Complexity*, 8(4), 31–38, 2003.

58. C. Zhang, J. Zhang, G. Xie, and L. Wang. Group penalty on the evolution of cooperation in spatial public goods games. *J. Stat. Mech.*, 12004, 2010, 2010.

59. D. Glätzle-Rützler, P. Lergetporer, S. Angerer, and M. Sutter. Third-party punishment increases cooperation in children through (misaligned) expectations and conditional cooperation. *Proc. Natl Acad. Sci. USA*, 111(19): 6916–6921, 2014.

60. R. Boyd, H. Gintis, S. Bowles, and P. J. Richerson. The evolution of altruistic punishment. *Proc. Natl Acad. Sci. USA*, 100:3531–3535, 2003.

61. A. Szolnoki, G. Szabó, and M. Perc. Phase diagrams for the spatial public goods game with pool punishment. *Phys. Rev. E*, 83:036101, 2011.

62. D. G. Rand, J. J. Armao, M. Nakamaru, and H. Ohtsuki. Anti-social punishment can prevent the co-evolution of punishment and cooperation, *J. Theor. Biol.*, 265:624–632, 2010.

63. L. Balafoutas, N. Nikiforakis, and B. Rockenbach. Altruistic punishment does not increase with the severity of norm violations in the field. *Nat. Commun.*, 7, 2016.

64. D. G. Rand and M. A. Nowak. The evolution of anti-social punishment in optional public goods games. *Nat. Commun.*, 2:434, 2011.

65. T. Chu, J. Zhang, and F. J. Weissing. Does insurance against punishment undermine cooperation in the evolution of public goods games? *J. Theor. Biol.*, 321:78–82, 2013.

66. D. Helbing, A. Szolnoki, M. Perc, and G. Szabó. Punish, but not too hard: how costly punishment spreads in the spatial public goods game. *New J. Phys.*, 12:083005, 2010.

67. E. Fehr and S. Gachter. Human behaviour: Egalitarian motive and altruistic punishment (reply). *Nature*, 433:E1–E1, 2005.

68. J. Zhang. *Coexistence of Competing Strategies in Evolutionary Games*. Rijksuniversiteit Groningen, Groningen, 2015.

# Chapter 3

## The Epidemic Spreading Processes in Complex Networks

Chengyi Xia, Zhishuang Wang, and Chunyun Zhen

**CONTENTS**

## 3.1   Introduction

Throughout the history of human society, the infectious diseases continually threaten the human life and health, from the Black Death (i.e., bubonic plague) of Europe in the Medieval century, the world-wide flu outbreaks starting from 1918, to the SARS, Ebola, and several emerging flus (e.g., H1N1, H5N1, H7N9) in the recent years [1–4]. Meanwhile, via the developing network communication techniques, the computer virus almost penetrates into each corner within the Internet and even endangers the cyber-space security [5–7]. Thus, how to analyze and understand the epidemic spreading behavior, and then try to predict and contain the epidemic outbreaks, has become a significant and tough issue for the global society [8–11].

Since it is not allowed to conduct the experimental studies within the population, at present, the mathematical modeling for the infectious disease spreading provides a powerful framework to perform the theoretical analysis and prediction for the disease diffusion behavior [1]. In particular, models that enable the control and suppression of epidemics have created the great interests within the academic communities, including the epidemiology, public health, sociology, mathematics, computer science, and even physics [12,13]. The first mathematical model was proposed by Bernoulli to investigate the spreading properties of smallpox [14], and then a wide number of mathematical models had been presented to depict the epidemic spreading characteristics, in which the population will be categorized into several compartments. Among them, the most famous disease models may be the SIS (Susceptible–Infected–Susceptible) and SIR (Susceptible–Infected–Removed) ones put forward by Kermack and McKendrick [15], where the individuals can only exist in the two or three distinct compartments, such as Susceptible (**S**), Infected or Infective (**I**), and Recovered or Removed (**R**). As an example, the smallpox and measles are often characterized with the SIR model as the individuals recovered from these diseases will not be infected again, while the seasonal influenza can be described with the SIS model because the susceptible ones can be infected more than once even if the infective agents can be cured.

Over the past centuries, the mathematical modeling literatures concerning the epidemic spreading often assume that the population is well mixed [1], that is, any individual can interact with another one with the identical probability within the population for the ease of analytical processing. However, it is far from the reality on account of the loss of population structure. In particular, two seminal works find that the small world [16] and scale-free [17] properties are often exhibited within many real-world systems which are represented as the complex networks, where the nodes stand for the interacting individuals and the links mimic the connection or interplay relationship among them. Thus, on the basis of complex network frameworks [18,19], exploring the spreading behavior within the real-world population or virus transmission on Internet has become an intriguing topic, and the role of contact structure between individuals (i.e., topology of contact network) cannot be ignored when the epidemic modeling is performed [20]. It is notable to mention that Pastor-Satorras and Vespignani [21] probed into the effect of contact network topology on the disease spreading for the first time by using complex networks, and found that the heterogeneous topology may lead to the absence of epidemic threshold under the thermodynamic limit, that is, the epidemics will be persistent once it emerges, and this surprising result dramatically differs from the conclusions obtained in the traditional mathematical epidemiology, where there is always a limited and positive critical threshold for epidemics [1,12,13].

After this pioneering work, large quantities of researches are devoted to resolving the role of population contact network structure in the epidemic spreading behavior, where the Monte Carlo (MC) simulations and theoretical studies are often combined to discuss the impact of heterogeneous topology on the critical threshold of epidemics, dynamical spreading characteristics and diffusion velocity, epidemic variability, and prevalence at the stationary state [20]. Among them, the epidemic threshold indicates whether the epidemics can be prevalent within the population or not, the epidemic will become endemic (i.e., there is a finite fraction of individuals who are finally infective) if the transmission probability is greater than the threshold, but the epidemic becomes extinct provided that the transmission probability is lower than this threshold. Near the epidemic threshold, the system may exhibit many interesting phenomena, such as rare-region phenomenon and the scaling behavior of the main magnitudes. Meanwhile, other macroscopic measures of complex networks, including the clustering properties [22], node or link weight distribution [23], community structure [24,25], hierarchical [26], and spatial structure [27], can also have a strong influence on the contagion behavior. Furthermore, several microscopic mechanisms, which include the saturation effect [28], degree correlated transmission [29], constant or nonlinear infectivity [30,31], piecewise infectivity [32], and so on, are introduced into the epidemic models in complex networks, and it is found that these schemes have rendered the positive threshold in complex networks with highly heterogeneous topology. Henceforth, it is significant for us to further investigate the specific infection mechanism to model the epidemic spreading on top of networks under some real scenarios.

In the meantime, many systems are often interconnected or interdependent [33], and hence the disease-related information may also diffuse within the same population as the epidemic spreads in the contact networks, where the information spreading topology can be distinct from the disease propagation network, and therefore, the two-layered network epidemic model has been constructed by Granell et al. [34] to characterize the interplay between the information (i.e., awareness) and disease, and it is indicated that the evolution of the epidemic threshold depends upon the topological structure of the multiplex networks and the interrelation between the awareness process and disease spreading, and then the meta-critical point regarding the epidemic onset has been induced by the information and its corresponding topology. However, this model assume that the nodes become immediately aware of the epidemics once they are infected, and meanwhile aware ones are totally immune to the infection in the future. By relaxing these two assumptions, they integrated the mass media that disseminates the diseased information within the whole population into the system, and found that the efficacy of vaccine and mass media can affect the critical relation between these two competing spreading processes, and then the meta-critical point disappears as a result of the introduction of mass media [35]. In addition, the social dynamics are further incorporated into the information diffusion model [36], where agents act like a herd effect and make a decision according to the status of their nearest neighbors, and it is clearly shown that a local awareness ration around 0.5 has a two-stage impact on the epidemic threshold and size of prevalence irrespective of the underlying network topology structure. Based on the framework of multiplex networks, the studies on the interaction between the disease and awareness are on its infancy and still worth deeply exploring in the future.

The remainder of this chapter is organized as follows. Firstly, in Section 3.2, we consider the role of nonuniform transmission and time delay from recovering process of infective individuals, and investigate the impact of combination of two different infection schemes on the epidemic threshold in complex networks. Secondly, we extend the epidemic model on two-layered multiplex network and propose a novel scheme where the SIR model is adopted in the disease spreading layer while the SIS-like two-state model is considered on the awareness layer, and the epidemic threshold for this model has also been analytically derived according to the microscopic Markov chain approach (MMCA) in Section 3.3. Finally, in Section 3.4, we end this chapter with some concluding remarks and point out the potential topics in the future.

## 3.2    Epidemic Spreading in Single-Layer Complex Networks

### 3.2.1    The novel disease spreading model in complex networks

As mentioned above, in the standard SIS model, individuals are divided into two categories: Susceptible (**S**) and Infected (**I**). Susceptible individuals are healthy ones that can be infected with the probability $\beta$ through contacts with infectious subjects. Infective individuals in their turn are recovered with the probability $\gamma$. Hence, individuals may go through the cycle $\mathbf{S} \rightarrow \mathbf{I} \rightarrow \mathbf{S}$ and their dynamics can be described by

$$
\begin{cases}
\dfrac{ds(t)}{dt} = -\gamma\rho(t) + \beta s(t)\rho(t) \\[2mm]
\dfrac{d\rho(t)}{dt} = \gamma\rho(t) - \beta s(t)\rho(t)
\end{cases}
\tag{3.1}
$$

where $s(t)$ and $\rho(t)$ stand for the fraction of susceptible and infective individuals. Generally, we neglect the details of disease infection and fix the size of the total population, and thus $s(t)$ and $\rho(t)$ need to satisfy the normalization condition: $s(t) + \rho(t) = 1$.

In this section, as published in Reference 37, we propose a modified SIS model with nonuniform spreading or transmission probabilities and delayed recovery, in which we still assume that individuals can be susceptible or infectious. However, we introduce two new ingredients:

- After an individual is infected by his/her infected neighbors at any time step $t$, it will be infectious during a time window $T + 1$. Once this time has elapsed, the infective agent goes back to the susceptible state, $S$, with the probability $\gamma = 1$, which can be assumed without loss of generality.

- At each time step $t$, infected individuals spread the disease to susceptible nodes with a probability that depends on the number of connections it has. Therefore, we assume that the effective spreading rate $\lambda = \frac{\beta}{\gamma}$ is a

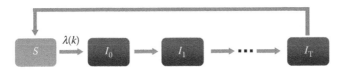

**FIGURE 3.1:** The schematic diagram of SIS epidemic model with nonuniform transmission and recovery delay. In this model, we assume that after the initial infection, the newly infected node will remain infectious during a time window of $T + 1$ time steps, after which the node recovers and gets back to the susceptible state.

degree-dependent function $\lambda(k) = \lambda_0 \frac{k^\alpha}{k}$ (i.e., so-called nonuniform transmission), where $k$ is the degree of an individual, $\lambda_0$ is an independent infection rate and $\alpha$ is a tunable parameter.

The disease spreading process for this modified model can be illustrated in Figure 3.1, in which $I_0, I_1, \ldots, I_T$ denote the infective individuals at different stages and $S$ represents the susceptible agents.

### 3.2.2 Mean-field theoretical analysis

In this subsection, we will investigate the epidemic thresholds of this modified model in both homogeneous and heterogeneous networks, respectively.

#### 3.2.2.1 Homogeneous topology

In homogeneous networks, the degree distribution is highly peaked around the average degree of the graph and the probability of finding a node with degree larger than the average degree $< k >$ decays exponentially fast as we move away from the peak. Making use of the mean-field approximation, we can assume the connectivity of each node to be constant and equal to $< k >$. In this case, $\lambda(k)$ also becomes constant for all the nodes and reads as $\lambda = \lambda' = \lambda_0 \frac{k^\alpha}{k}$. Now, we can write the dynamical equations describing the system dynamics as:

$$
\begin{cases}
\dfrac{\partial s(t)}{\partial t} = -\lambda' < k > s(t)\rho(t) + \rho_T(t) \\[2mm]
\dfrac{\partial \rho_0(t)}{\partial t} = -\rho_0(t) + \lambda' < k > s(t)\rho(t) \\[2mm]
\dfrac{\partial \rho_1(t)}{\partial t} = -\rho_1(t) + \rho_0(t) \\[2mm]
\quad \cdots \\[2mm]
\dfrac{\partial \rho_T(t)}{\partial t} = -\rho_T(t) + \rho_{T-1}(t)
\end{cases}
\tag{3.2}
$$

where $s(t)$ is the fraction of susceptible individuals, $\rho_0(t), \ldots, \rho_T(t)$ denote the fraction of individuals in the states $I_0, \ldots, I_T$, respectively, and $\rho(t) = \sum_{j=0}^{T} \rho_j(t)$. The first equation in Equation 3.2 represents the time evolution of susceptible individuals while the rest of the equations stand for the evolution of each of $T$ stages through which infected individuals pass until the final recovery. In the first equation, the negative term accounts for the new contagions and the positive term stands for the recovery of infected individuals. In the rest of Equation 3.2, each negative term represents nodes passing to the next stage of the recovery time while the positive one includes the passage from the previous stage to the present one. If we insert the normalization condition $s(t) + \rho(t) = 1$ into Equation 3.2, we get

$$\frac{\partial \rho_0(t)}{\partial t} = -\rho_0(t) + \lambda' < k > [1 - \rho(t)]\rho(t). \tag{3.3}$$

When $t \to \infty$, the system reaches a stationary state in which $\frac{\partial \rho_j(t)}{\partial t} = 0$ for $j = 0, 1, \ldots, T$ and hence,

$$\rho_0 = \rho_1 = \ldots = \rho_T, \tag{3.4}$$

where $\rho_0, \rho_1, \ldots, \rho_T$ is used to denote the steady state values of $\rho_0(t), \rho_1(t), \ldots, \rho_T(t)$, respectively. Likewise, we make use of $\rho$ to stand for the value of $\rho(t)$ at the stationary state. Finally, we have the following equality,

$$\rho_0 = \rho_1 = \ldots = \rho_T = \frac{\rho}{T+1}, \tag{3.5}$$

Combining Equations 3.2 and 3.5, one obtains

$$-\lambda' < k > [1 - \rho]\rho + \frac{\rho}{T+1} = 0 \tag{3.6}$$

Obviously, $\rho = 0$ is a trivial solution to Equation 3.6, which corresponds to the disease-free state. To obtain a nontrivial solution $0 < \rho < 1$, which would represent an endemic state, the following condition must be fulfilled,

$$\frac{d}{dt}[\lambda' < k > (T+1)(\rho - \rho^2)]|_{\rho=0} = 1 \tag{3.7}$$

That is,

$$\lambda' < k > (T+1) = 1 \tag{3.8}$$

At variance with the standard formulation, we assume that the infectivity $\lambda'$ decreases with the number of neighbors of each individual (i.e., the degree $k$

which is equal to the average degree of population $<k>$), verifying $\lambda' = \frac{\lambda_0 k^\alpha}{k}$ in which $0 \leq \alpha \leq 1$ and $\lambda_0$ is a degree-independent spreading rate. This expression, for the case of homogeneous networks, reduces to $\lambda' = \frac{\lambda_0 <k>^\alpha}{<k>}$ for all individuals. Thus, the critical threshold can be written as:

$$\lambda_0^c = \frac{1}{(T+1) <k>^\alpha} \tag{3.9}$$

Although Equation 3.9 means the general values of the epidemic threshold, some particular cases are worth being further discussed as follows:

- When the recovery delay is absent (i.e., $T=0$), the critical threshold is governed by the nonuniform transmission $\lambda_0^c = \frac{1}{<k>^\alpha}$. In particular, $\alpha = 1$ stands for the standard case where the critical spreading rate is equal to the inverse of the average degree of the network, at which we recover the classical result $\lambda_0^c = \frac{1}{<k>}$ [20,21].

- When $\alpha = 0$, if we introduce a re-scaling factor $A$ at the pre-factor of $\lambda'$, we get $\lambda' = A\lambda_0 k^{-1}$. In this case, the critical threshold reads as $\lambda_0^c = \frac{1}{T+1}\frac{1}{A}$, and therefore, the threshold is independent of the network topology. This re-scaling leads, in the particular case of $T=0$, to be $\lambda_0^c = \frac{1}{A}$ [30].

### 3.2.2.2 Heterogeneous topology

As pointed out in the previous works [18,19], for many natural, social, and engineering systems, the degree distribution is highly skewed, that is, the topology is very heterogeneous and the average degree is not anymore a good proxy for the degree of each individual. In this case, nonuniform transmission effects are expected to be much more relevant in terms of their influence on the spreading dynamics, as the degree dependence of the spreading rate $\lambda(k) = \lambda_0 \frac{k^\alpha}{k}$ is explicitly incorporated. Hence, we should write the distinct equations for each class of degree $k$, which read as

$$\begin{cases} \dfrac{\partial s_k(t)}{\partial t} = -k s_k(t)\Theta_k(t) + \rho_{k,T}(t) \\[2mm] \dfrac{\partial \rho_{k,0}(t)}{\partial t} = -\rho_{k,0}(t) + k s_k(t)\Theta(t) \\[2mm] \dfrac{\partial \rho_{k,1}(t)}{\partial t} = -\rho_{k,1}(t) + \rho_{k,0}(t) \\[2mm] \cdots \\[2mm] \dfrac{\partial \rho_{k,T}(t)}{\partial t} = -\rho_{k,T}(t) + \rho_{k,T-1}(t) \end{cases} \tag{3.10}$$

where $\rho_{k,T}(t)$ denotes the fraction of individuals with degree $k$ in state $I_T$ at time step $t$, and $\Theta_k(t)$ is the probability that a randomly chosen link points to an infected individual, and $\Theta_k(t)$ is generally labeled as,

$$\Theta_k(t) = \sum_{k'} \lambda(k) P(k'|k) \rho_{k'}(t), \qquad (3.11)$$

where $\rho_{k'}(t)$ is the density of infected individuals with degree $k$ and the conditional probability $P(k|k)$ denotes the probability that a node with degree $k$ is connected to a node with degree $k'$. We assume that networks are uncorrelated, it is impled that $\Theta_k(t) = \Theta_{k'}(t) = \Theta(t)$ and $P(k|k) = \frac{kP(k)}{\sum_s sP(s)}$, which further gives out the following equality,

$$\Theta(t) = \sum_{k'} \frac{\lambda(k')k'P(k')\rho_{k'}(t)}{\sum_s sP(s)} = \frac{\sum_{k'} \lambda(k')k'P(k')\rho_{k'}(t)}{<k>}. \qquad (3.12)$$

At the steady state, $\frac{\partial \rho_{k,\tau}(t)}{\partial t} = 0$ for $\tau = 0, 1, \dots, T$, and let $\rho_k(t) = \sum_{\tau=0}^{T} \rho_{k,\tau}(t)$ be the density of infected individuals with degree $k$, with the steady state values of $\rho_k(t), \rho_{k,0}(t), \dots, \rho_{k,T}(t)$ being labeled as $\rho_k, \rho_{k,0}, \dots, \rho_{k,T}$, respectively. Equation 3.10 can be reduced as:

$$\begin{cases} -\rho_{k,0} + k(1-\rho_k)\Theta = 0 \\ \qquad -\rho_{k,1} + \rho_{k,0} = 0 \\ \qquad\qquad \dots \\ -\rho_{k,T} + \rho_{k,T-1} = 0 \end{cases} \qquad (3.13)$$

Inserting $\rho_k = \sum_{\tau=0}^{T} \rho_{k,\tau}$ into Equation 3.13, we can obtain

$$\begin{cases} \rho_{k,0} = \rho_{k,1} = \dots = \rho_{k,T} = \dfrac{\rho_k}{T+1} \\[2ex] \rho_k = \dfrac{k\Theta}{\dfrac{1}{T+1} + k\Theta} \end{cases} \qquad (3.14)$$

which then gives the relationship as follows:

$$\Theta = \frac{1}{<k>} \sum_{k'} \lambda(k')k'P(k') \frac{k'\Theta}{\dfrac{1}{T+1} + k'\Theta} \qquad (3.15)$$

From the above equation, it is obvious that $\Theta = 0$ is a trivial solution to Equation 3.15 which corresponds to the absence of an outbreak. In order to have a nontrivial solution, $0 < \Theta < 1$, the following condition must be satisfied,

$$\frac{d}{d\Theta}\left[\frac{1}{<k>}\sum_{k'}\lambda(k')k'P(k')\frac{k'\Theta}{\frac{1}{T+1}+k'\Theta}\right]\Bigg|_{\Theta=0} > 1. \qquad (3.16)$$

which means that

$$\frac{1}{<k>}\sum_{k'}\lambda(k')k'P(k')k'(T+1) > 1 \qquad (3.17)$$

Once again, according to Equation 3.17, there are different scenarios worth being remarked as follows:

- If we assume that the nonuniform transmission effect is absent and so the spreading rate is independent of the degree of the nodes (i.e., $\alpha = 1$), then $\lambda(k) = \lambda_0 \frac{k}{k} = \lambda_0$, and Equation 3.17 reads as:

$$\lambda_0 > \frac{1}{T+1} \cdot \frac{<k>}{<k^2>} \qquad (3.18)$$

  That is, the threshold $\lambda_0^c = \frac{1}{T+1} \cdot \frac{<k>}{<k^2>}$. In particular, $T = 0$ gives out the threshold $\lambda_0^c = \frac{<k>}{<k^2>}$ which recovers the classical conclusion in References 20 and 21.

- If the spreading rate is proportional to the inverse of the degree (i.e., $\alpha = 0$ denoting the identical nonuniform transmission), we recover the expression previously obtained for the homogeneous network by considering the pre-factor $A$, i.e.,

$$\lambda_0^c > \frac{1}{T+1} \cdot \frac{1}{A} \qquad (3.19)$$

- For the intermediate cases, i.e., $\lambda(k) = \lambda_0 \frac{k^\alpha}{k}$ with $0 < \alpha < 1$, the critical threshold can be represented as

$$\lambda_0^c > \frac{1}{T+1} \cdot \frac{<k^{1+\alpha}>}{k} \qquad (3.20)$$

### 3.2.3  Numerical simulations

So as to validate the mean-field results, we next perform large-scale numerical simulations on ER-like random networks, which are generated using the Watts–Strogatz (WS) model with the rewiring probability $p = 1$, and BA scale-free networks. All simulation results are averaged over 100 independent runs and, unless otherwise stated, the size of the networks is $N = 5000$.

Moreover, we set the average degree of both kinds of graphs to be $<k> = 8$ and simulations have been done using a recovery rate $\gamma = 1$.

Figure 3.2 shows the steady state density of infected individuals or prevalence $\rho(\infty)$ as a function of the effective spreading rate $\lambda_0$ in both networks. Here we do not consider any recovery delay, that is, $T = 0$. In this figure, the arrows

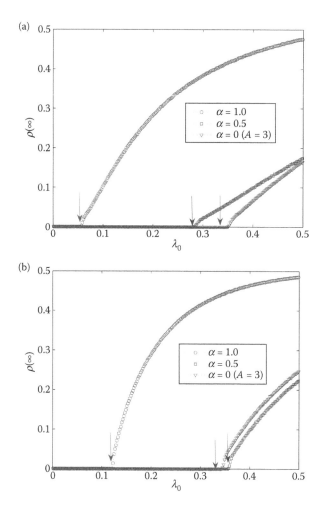

**FIGURE 3.2:** The effect of degree-dependent spreading rates on the propagation of diseases in complex networks. In this case, the delayed recovery mechanism is not present. (a) BA scale-free model. (b) Random WS model ($p = 1$). All data points are obtained after averaging 100 independent runs and the dashed lines are a guide to the eyes. The arrows denote the epidemic thresholds as given by the analytical results. The parameter $A$ in these figures plays the role of a scaling factor for the critical point. When $A = 1$, the contact process is recovered for the case $\alpha = 0$ and so the critical point is $\lambda_0^c = 1$.

correspond to the analytical results for the epidemic thresholds for several values of $\alpha$ as indicated. In most cases, the analytical predictions agree well with the results of numerical simulations. However, for $\alpha = 0$ and $A = 3$, the theoretical results are slightly different from the numerical ones, which we think is due to finite size effects. In any case, the figure nicely shows that the critical thresholds for an outbreak to occur can be largely altered by this nonuniform transmission scheme.

On the other hand, Figure 3.3 illustrates the effects of the second mechanism— the delayed recovery—when it is the only ingredient at work with respect to the

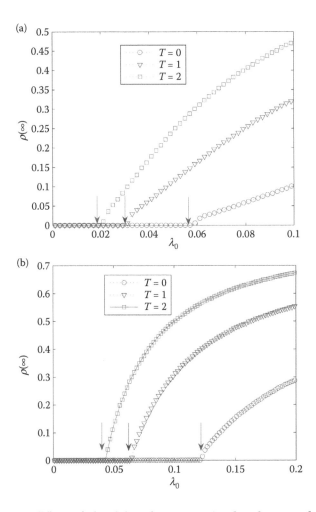

**FIGURE 3.3:** Effect of the delayed recovery in the absence of the degree-dependent spreading rates scenario (i.e., $\alpha = 1.0$). (a) BA scale-free model. (b) Random WS model ($p = 1$). All data points are obtained after averaging 100 independent runs and the dashed lines are a guide to the eyes. The arrows denote the epidemic thresholds as given by the analytical results.

standard case. Again, the results from numerical simulations are in accordance with our mean-field calculations and confirm that if infected individuals remain so for longer times, the critical threshold is largely reduced.

Another scenario of interest corresponds to the case in which the spreading rate is inversely proportional to the degree of each individual (i.e., $\lambda = \lambda_0 k^{-1}$ which means that $\alpha = 0$). As it can be shown, in this case, in what regards the epidemic thresholds whether the network is homogeneous or heterogeneous is completely irrelevant, provided that both kinds of graphs have the same average degree $<k>$. Figure 3.4 shows the numerical results, which confirm that the

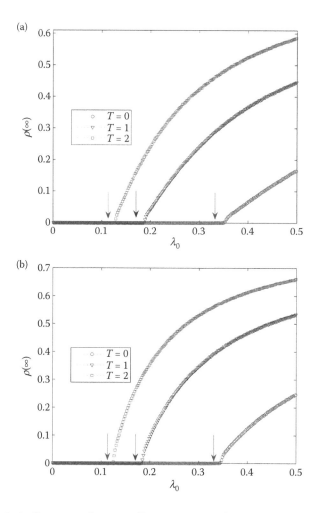

**FIGURE 3.4:** Same results as in Figure 3.3 but for $\alpha = 0$ and $A = 3$. (a) BA scale-free model. (b) Random WS model ($p = 1$). All data points are obtained after averaging 100 independent runs and the dashed lines are a guide to the eyes. The arrows denote the epidemic thresholds as given by the analytical results.

underlying networks do not determine the epidemic thresholds. Actually, they are the same in both panels for all values of $T$ simulated.

Figure 3.5 describes what happens when both mechanisms are simultaneously active. Essentially, a delay in recovery will make the epidemic threshold smaller than for the standard case and unless the dependence of the spreading rate compensates for this effect, the disease-free regime will be shortened, as it happens after our parameter selection in this case.

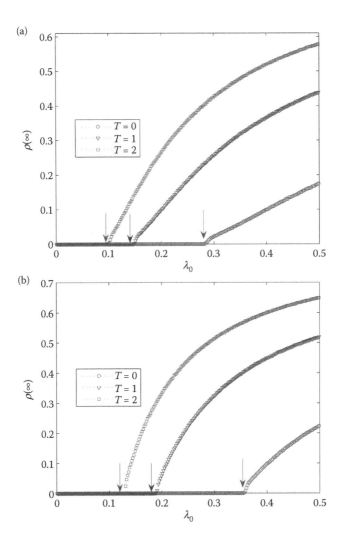

**FIGURE 3.5:** Numerical results when both mechanisms studied are concurrently active. (a) BA scale-free model. (b) Random WS model ($p=1$). The value of $\alpha$ has been set to 0.5. Arrows identify the epidemic thresholds as given by the mean-field approach.

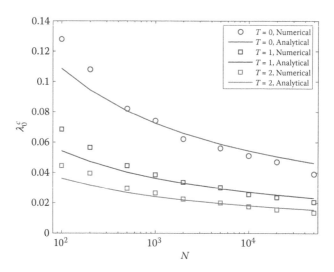

**FIGURE 3.6:** Comparison of critical threshold between the numerical simulations and analytical results in Equation 3.20 in which $\alpha$ is fixed to be 1.0 and the network parameters are set to be $m = m_0 = 4$ in BA model.

Finally, with the aim to clarify the nature of the small discrepancies between numerical results and the analytical predictions of Equation 3.20, in Figure 3.6, we consider the influence of network size on the epidemic threshold. Results highlight that, as expected, the critical threshold decreases as the network size increases. Simultaneously, the mismatch between numerical and theoretical values drops, suggesting that the initial differences are due to finite size effects.

## 3.3    Epidemic Spreading in Multi-Layer Complex Networks

### 3.3.1    The novel spreading model in two-layered multiplex networks

In order to further model the interplay between the disease propagation and its related information, similar to References 34 and 35, we select a two-layered multiplex network, shown in Figure 3.7, as the underlying framework for the evolution of our dynamical systems. Here, one layer (i.e., Lower layer) represents the physical contact network for the epidemic spreading, and the other one (i.e., Upper layer) is used for the virtual communication network for the information diffusion related with epidemics. Although the mapping mode between nodes for these two networks is one to one, the linking relationship on one layer is different from that on the other one, that is, the topology between two-layer networks is distinct. In addition, these two layers are assumed to be unweighted and undirected for the convenience of analytical processing, and

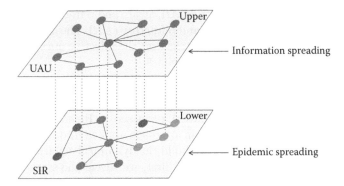

**FIGURE 3.7:** The schematic diagram of multiplex networks used in the current UAU–SIR model. The upper layer represents the information spreading, where each node can only exist one of two possible states: unaware (U, dark nodes) and aware (A, light nodes). The lower layer stands for the epidemic spreading with three compartments: susceptible (S, dark nodes), infective (I, light dark nodes), and recovered (R, light nodes). Combining the individual states between two layers, only five possible states have been considered in this work, that is, the unaware and infective (UI) state is excluded. The coupled dynamics between two layers iterates continuously as the discrete time steps evolves.

the one-to-one mapping among nodes corresponds to the dynamical interrelation between two layers.

For the information propagation on the upper layer, the SIS-style two-state contact process is adopted here, where an individual may lie in the aware (A) state if he is aware of epidemics, or else he will be in the unaware (U) state. While for the epidemic contagion on the lower layer, unlike many other works [34,36, 38], we adopt the SIR model as the basis of disease evolution, in which individuals can be categorized into three compartments: susceptible (S), infected (I), and recovered (R). Regarding the information diffusion, unaware individuals switch to the aware state with the probability $\lambda$ due to the following two reasons: one is that they have been infected, the other one is that one of their nearest neighbors has been in the aware state and informed them; an aware individual becomes unaware with the probability $\delta$ because he has forgotten the current information on epidemic state, and he has become re-susceptible or kept susceptible unchanged. With regard to the contagion dynamics, like the classical SIR model, a susceptible individual will be changed into the infective state with the probability $\beta$ by means of contacting the infective neighbor, and meanwhile the infected ones will be recovered spontaneously with the curing rate $\mu$. Moreover, if an agent is infected, it is natural to assume that this individual becomes aware of the epidemics, but the infectivity of aware ones may be reduced by an attenuation factor $0 \leq \gamma \leq 1.0$. Here, $\beta^A = \gamma\beta^U = \gamma\beta$ and $\beta^U = \beta$ are used to denote the infection rates with and without awareness, respectively. Henceforth, each individual within this model can only have five potential states: aware and susceptible (*AS*), aware and infective (*AI*), aware and refractory (*AR*), unaware and susceptible (*US*), and unaware and refractory (*UR*). The unaware and infective (*UI*) state is excluded as we suppose that the infective ones certainly know the related information about the epidemics.

### 3.3.2   MMCA theoretical analysis

In this section, resorting to MMCA proposed in Reference 39, which is a variant of discrete-time Markov chain approach applied into the analysis of epidemics in complex networks, we will build the dynamical equation of five possible states for the UAU–SIR model. At first, we reveal the state transmission relationship through the probability tree method, illustrated in Figure 3.8.

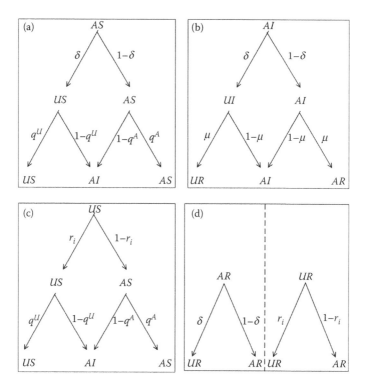

**FIGURE 3.8:** The probability transmission trees for five possible states including *AS*, *AI*, *AR*, *US*, and *UR*. We denote the awareness forgotten (from aware to unaware state) probability with $\delta$, and $r_i$ means the probability not varying from unaware to aware state. $\mu$ means the probability for an infective individual to become recovered, $q^A$ represents the transition probability for an individual not being infected by his neighbors if he is aware, and $q^U$ denotes the transition probability for an individual not being infected by his neighbors if he is unaware. The state transition process proceeds alternately between information and disease spreading within each time step.

Let $A = (a_{ij})$, $B = (b_{ij})$ denote the adjacent matrices of upper- and lower-layer networks, respectively. At each time step $t$, an individual $i$ can be only in one of five possible states as defined in Section 3.2, and their corresponding fractions are represented as $p_i^{AS}(t)$, $p_i^{AI}(t)$, $p_i^{AR}(t)$, $p_i^{US}(t)$, and $p_i^{UR}(t)$. After that, on the upper layer, we define the probability for an unaware individual $i$ not to enter into the aware state as $r_i(t)$; on the lower layer, $q_i^A(t)$ and $q_i^U(t)$ represent the probabilities for aware and unaware susceptible individual not being infected by any infective neighbors, respectively. Based on these definitions, we have the following formulae

$$
\begin{cases}
r_i(t) = \displaystyle\prod_j \left[1 - a_{ji}p_j^A(t)\lambda\right] \\[2ex]
q_i^A = \displaystyle\prod_j \left[1 - b_{ji}p_j^{AI}(t)\beta^A\right] \\[2ex]
q_i^U = \displaystyle\prod_j \left[1 - b_{ji}p_j^{AI}(t)\beta^U\right]
\end{cases}
\tag{3.21}
$$

where $j$ runs over all nearest neighbors of individual $i$, and $a_{ji}$ and $b_{ji}$ are set to be 1 if $j$ is connected to $i$, or else they are zero.

Then, according to the state transmission in Figure 3.8, we build the dynamically evolutionary equations for five possible states by using the MMCA method as follows

$$
\begin{cases}
p_i^{US}(t+1) = p_i^{US}(t)r_i(t)q_i^U(t) + p_i^{AS}(t)\delta q_i^U(t) \\[1.5ex]
p_i^{AS}(t+1) = p_i^{US}(t)[1 - r_i(t)]q_i^A(t) + p_i^{AS}(t)(1-\delta)q_i^A(t) \\[1.5ex]
p_i^{AI}(t+1) = p_i^{AI}(t)(1-\mu) \\[1ex]
\qquad\quad + p_i^{AS}(t)\{[1 - q_i^A(t)](1-\delta) + [1 - q_i^U(t)]\delta\} \\[1ex]
\qquad\quad + p_i^{US}(t)\{[1 - r_i(t)][1 - q_i^A(t)] + r_i(t)[1 - q_i^U(t)]\} \\[1.5ex]
p_i^{AR}(t+1) = p_i^{AI}(t)(1-\delta)\mu + p_i^{AR}(t)(1-\delta) + p_i^{UR}(t)[1 - r_i(t)] \\[1.5ex]
p_i^{UR}(t+1) = p_i^{AI}(t)\delta\mu + p_i^{AR}(t)\delta + p_i^{UR}(t)r_i(t)
\end{cases}
\tag{3.22}
$$

where $p_i^{US}(t)$ or $p_i^{US}(t+1)$ in the first equation denote the fraction of individuals who are in the U.S. state at the current time step $(t)$ or next time step $(t+1)$, the parallel quantities (e.g., $p_i^{AS}$, $p_i^{AI}$, $p_i^{AR}$, $p_i^{UR}$) in other equations have the similar denotations, meanwhile they need to satisfy the normalized condition: $p_i^{AS} + p_i^{AI} + p_i^{AR} + p_i^{US} + p_i^{UR} = 1$. In particular, when $t \to \infty$, the system arrives at the stationary state and we can assume that $p_i^{AI}(t+1) = p_i^{AI}(t) = p_i^{AI}$

is true for the AI state, and the same relationship also holds for other four quantities.

As the epidemic threshold is the most vital quantity in the clinic diagnosis and public hygienics, which will determine whether the epidemics finally outbreak or die out, estimating the critical threshold for infectivity $(\beta_c = \beta_c^U)$ on the contagion layer is the first step to analyze the characteristics of the current model. Around the threshold, the probability of being infected for any individual approaches zero and we can assume that $p_i^{AI} = \epsilon_i \ll 1$. Accordingly, starting from the above assumption and Equation 3.21, the probabilities for individuals not being infected by any infective neighbor can then be written as

$$
\begin{cases}
q_i^A = \prod_j \left[ 1 - b_{ji} p_j^{AI}(t) \beta^A \right] \\[4pt]
\quad \approx \left( 1 - \beta^A \sum_j b_{ji} \epsilon_j \right) \\[4pt]
q_i^U = \prod_j \left[ 1 - b_{ji} p_j^{AI}(t) \beta^U \right] \\[4pt]
\quad \approx \left( 1 - \beta^U \sum_j b_{ji} \epsilon_j \right)
\end{cases}
\tag{3.23}
$$

where the higher-order items on $\epsilon_j$ have been neglected since $\epsilon_j$ is small enough near the epidemic threshold. For the brevity, let $\alpha^A = \beta^A \sum_j b_{ji} \epsilon_j$, $\alpha^U = \beta^U \sum_j b_{ji} \epsilon_j$, with the help of Equation 3.23, we can rewrite the first three equations of Equation 3.22 as follows

$$
\begin{cases}
p_i^{US}(t+1) = p_i^{US}(t) r_i(t)(1 - \alpha^U) + p_i^{AS}(t) \delta(1 - \alpha^U) \\[4pt]
p_i^{AS}(t+1) = p_i^{US}(t)[1 - r_i(t)](1 - \alpha^A) + p_i^{AS}(t)(1 - \delta)(1 - \alpha^A) \\[4pt]
p_i^{AI}(t+1) = p_i^{AI}(t)(1 - \mu) \\[4pt]
\qquad + p_i^{AS}(t)\{\alpha^A(1 - \delta) + \alpha^U \delta\} \\[4pt]
\qquad + p_i^{US}(t)\{[1 - r_i(t)]\alpha^A + r_i(t)\alpha^U\}
\end{cases}
\tag{3.24}
$$

When $t \to \infty$, the stationary fractions of three states (including $p_i^{US}, p_i^{AS}, p_i^{AI}$) can be further reduced into the following relationship by ignoring all higher-order items in Equation 3.24,

$$
\begin{cases}
p_i^{US} = p_i^{US} r_i + p_i^{AS} \delta \\
p_i^{AS} = p_i^{US}(1 - r_i) + p_i^{AS}(1 - \delta) \\
\epsilon_i = \epsilon_i(1 - \mu) \\
\quad + p_i^{AS}[\alpha^A(1 - \delta) + \alpha^U \delta] \\
\quad + p_i^{US}[\alpha^A(1 - r_i) + \alpha^U r_i]
\end{cases}
\tag{3.25}
$$

As a further step, we can obtain

$$
\begin{aligned}
\epsilon_i &= \epsilon_i(1 - \mu) + \left[p_i^{AS}(1 - \delta) + p_i^{US}(1 - r_i)\right]\alpha^A + \left[p_i^{AS}\delta + p_i^{US}r_i\right]\alpha^U \\
&= \epsilon_i(1 - \mu) + p_i^{AS}\alpha^A + p_i^{US}\alpha^U \\
&= \epsilon_i(1 - \mu) + \left[p_i^{AS}\gamma + p_i^{US}\right]\beta^U \sum_j b_{ji}\epsilon_j
\end{aligned}
\tag{3.26}
$$

Let $p_i^{AS} + p_i^{AI} + p_i^{AR} = p_i^A$, meanwhile the normalization condition $p_i^{AS} + p_i^{AI} + p_i^{AR} + p_i^{UR} + p_i^{US} = 1$ needs to be satisfied. Furthermore, near the epidemic threshold, $p_i^{AI} \to 0$ and also $p_i^{UR} \to 0$, $p_i^{AR} \to 0$ provided that the initial fraction of infective agents is lower. Hence, we assume $p_i^{AS} \approx p_i^A$ and $p_i^{US} \approx 1 - p_i^A$, and the third equation of Equation 3.26 can be further simplified as

$$
\begin{aligned}
\mu\epsilon_i &= \left[p_i^A\gamma + (1 - p_i^A)\right]\beta^U \sum_j b_{ji}\epsilon_j \\
&= \beta^U\left[1 - (1 - \gamma)p_i^A\right] \sum_j b_{ji}\epsilon_j
\end{aligned}
\tag{3.27}
$$

Henceforth, the above equation can be further reduced to $\sum_j \{[1 - (1 - \gamma)p_i^A]b_{ji} - \frac{\mu}{\beta^U} t_{ji}\}\epsilon_j = 0$, in which $t_{ji}$ is the element of the identity matrix. Thus, the critical threshold $\beta_c = \beta_c^U$ can be considered as the solution of eigenvalue problems, which is the minimum value $\beta^U$ satisfying Equation 3.27. Let $\Lambda_{max}$ denote the maximal eigenvalue of $H$ whose elements are $h_{ji} = [1 - (1 - \gamma)p_i^A]b_{ji}$, and then the epidemic threshold regarding $\beta^U$ of the current two-layered model can be described as follows

$$
\beta_c^U = \frac{\mu}{\Lambda_{max}}
\tag{3.28}
$$

where $\mu$ denotes the recovering rate for infective individuals and $\Lambda_{max}$ is the maximal eigenvalue of matrix $H$. It is obvious that $\beta_c^U$ will be correlated with the spreading dynamics on the information layer, in particular, involving the values of $p_i^A$, and Equation 3.28 is totally similar to the results of UAU–SIS in Reference 34. In reality, the epidemic threshold on a traditional complex network (i.e., single-layer network) is also identical for the standard SIS and SIR model [21,40]. From the perspective of contagion dynamics, the epidemic threshold represents the critical value of the infection rate, beyond which the epidemics can be widely spread over the population or under which the epidemics will be extinct, and thus this threshold has become a fundamental measure to characterize the infectivity of infectious diseases in the public hygienic management and mathematical biology modeling.

### 3.3.3    Numerical simulations

As the analytical solution regarding the outbreak threshold in Section 3.2 is obtained by the MMCA, and there exists a moderate approximation during the process of problem conquest. Thus, we perform extensive MC simulations to check the accuracy of MMCA and then validate the comparisons between our analytical results and simulations from the coupled dynamics of our UAU–SIR model.

At first, we illustrate the fraction of recovered individuals $(\rho^R)$ on the contagion layer and aware ones $(\rho^A)$ for the information layer at the steady state as the spreading rate $\beta$ varies in Figure 3.9 for a fixed awareness spreading rate $\lambda$. Here the multiplex network has 10,000 nodes on each layer, but the scale-free topology with the power exponent 3.0 is created on the upper layer and the power exponent of scale-free topology on the lower layer is 2.5. For the MC simulations, we count the number of refractory individuals $(N_i^R)$ on the lower layer or aware ones $(N_i^A)$ on the upper layer at the stationary state, and then compute their fractions within the whole population as $\rho^R = \frac{\sum N_i^R}{N}$ and $\rho^A = \frac{\sum N_i^A}{N}$; while for the MMCA, we will record the iteration results at the last time step in the Equation 3.22. Here, the solid circles denote the MC simulations and the open diamonds represent MMCA, and it can be clearly observed that the results of MMCA qualitatively agree with those of MC simulations. On the one hand, when the attenuation factor of the spreading rate $\gamma$ is 0, that is, the aware individuals will be conferred to the complete immunity, the relative error regarding the refractory $\rho^R$ obtained from these two methods has arrived at 4.5%. However, when $\gamma$ is set to be 0.5, this error has been greatly reduced and approaches around 0.38% under this scenario, hence the MMCA has acquired a very high accuracy when compared to the MC simulations. On the other hand, when we compared the fraction of aware individuals $\rho^A$ on the upper layer, we can find that the consistency between two methods is excellent; in particular, the the spreading rate $\beta$ is beyond $\beta_c$, the disease outbreaks and the infective agents are finally recovered, accordingly the information spreading will not be correlated with the disease at the stationary state and the final $\rho^A$ will be solely determined by the information diffusion rate $\lambda$.

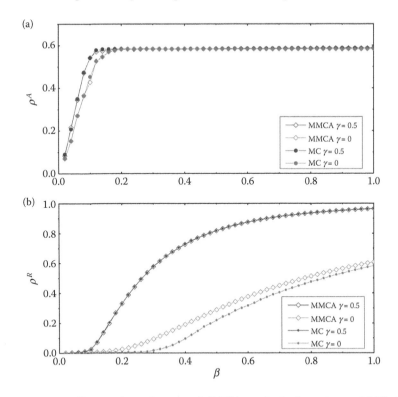

**FIGURE 3.9:** Comparisons between MMCA analytical results and MC simulations as the epidemic spreading rate $\beta$ increases in our UAU–SIR model. In panel (a), the fraction of aware nodes $\rho^A$ from MMCA and MC simulation is plotted as a function of infection rate $\beta$, where the open diamonds denote the results of MMCA with a different infectivity reduction factor $\gamma = 0$ (dark diamond) and $\gamma = 0.5$ (light diamond), while the solid circles represent those created by MC simulations from different $\gamma = 0$ (dark diamonds) and $\gamma = 0.5$ (light diamonds). In panel (b), the fraction of recovered individuals $\rho^R$ within the population with the abovementioned two methods is recorded to characterize the prevalence of disease spreading, likewise, the identical symbols have been adopted here. All MC simulations are averaged over 200 independent runs. The infective seeds have only 20 nodes, awareness spread rate $\lambda$ is fixed to be 0.8, and awareness loss rate is $\delta = 0.3$ and the curing rate for infective nodes is $\mu = 0.8$. Both layers are scale-free networks which consist of 10,000 nodes and have the power exponent 2.5 on the contagion layer and 3.0 on the information layer.

Therefore, it follows that MMCA approach has a well-defined performance and can mimic the coupled spreading dynamics in the case of multiplex networks.

Next, in Figure 3.10, we further explore the relationship between $\rho^R$ or $\rho^A$ and the information spreading rate $\lambda$ for a fixed transmission rate $\beta$ on the

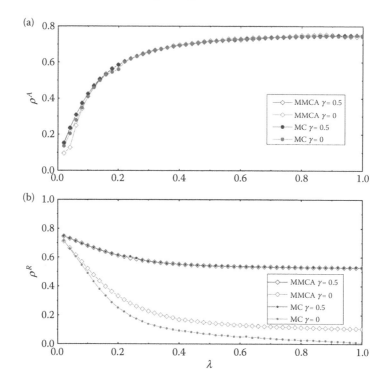

**FIGURE 3.10:** Comparisons between MMCA analytical results and MC simulations as the information spreading rate $\lambda$ varies in our UAU–SIR model. In panel (a), the fraction of aware nodes $\rho^A$ from MMCA and MC simulation is plotted as a function of infection rate $\lambda$, where the open diamonds denote the results of MMCA with a different infectivity reduction factor $\gamma = 0$ (dark diamond) and $\gamma = 0.5$ (light diamond), while the solid circles represent those created by MC simulations from different $\gamma = 0$ (dark circle) and $\gamma = 0.5$ (light circle). In panel (b), the fraction of recovered individuals $\rho^R$ within the population with the abovementioned two methods is recorded to characterize the prevalence of disease spreading as $\lambda$ increases, likewise, the identical symbols have been used here. All MC simulations are averaged over 200 independent runs. The infective seeds are still 20 nodes, the disease spreading rate $\beta$ is fixed to be 0.3, and awareness loss rate is $\delta = 0.3$ and the curing rate for infective nodes is $\mu = 0.8$. Both layers are scale-free networks which consist of 10,000 nodes and have the power exponent 2.5 on the contagion layer and 3.0 on the information layer.

contagion layer. As $\lambda$ increases, the individuals can be aware of the epidemics through the disease-related information diffusion, and thus $\rho^A$ will be gradually augmented, as shown in panel (a). Meanwhile, the infection risk for aware ones will be reduced and accordingly $\rho^R$ will be decreased, which can be observed in panel (b). However, the total immunity for aware ones, under which the correlation between two layers has created the extreme interaction, leads to the larger relative errors between MMCA and MC simulations on the contagion layer, which can arrive at 8.5% for $\gamma = 0$. When $\gamma = 0.5$, the interplay between the information diffusion and epidemic transmission will be moderate and MMCA can accurately approach the MC simulation results as the relative error between them has only 0.7%. While on the information layer, $\rho^A$ will be mainly determined by the information spreading rate $\lambda$ since the infective agents are absent in the final stage of contagion outbreaks, and $\rho^A$ is almost identical for MMCA and MC simulations whether $\gamma$ is 0 or 0.5. Again, the results obtained by the MMCA method can well agree with those created by the MC simulations, and MMCA can accurately forecast the disease dynamics coupled with the awareness transmission under the framework of multiplex networks.

In order to fully compare the results of MMCA and MC simulations with regard to the fraction of recovered nodes ($\rho^R$) at the stationary state, we plot the phase diagram ($\lambda - \beta$) for the coupled dynamics of UAU–SIR model, which has been conducted on the same multiplex networks as those in Figures 3.9 and 3.10. In Figure 3.11, the infectivity reduction coefficient $\gamma$ is set to be 0.5, total 2500 points in the whole plane are recorded, that is, 2500 (50 × 50) combinations for $\lambda - \beta$ are considered here, panel (a) presents the results obtained from MMCA while panel (b) depicts those computed with MC simulations. In the whole range of $\lambda - \beta$ parameters, we can find the good agreement for $\rho^R$ derived by these two methods, and the relative errors between them is less than 1% (around 0.34%). Provided that $\gamma$ is assumed to be 0, however, the quantitative error between them becomes a little more and arrived at up to 5% (at most 5.1%), in which the corresponding results are pictured in Figure 3.12. In short, the MMCA can well approximate the coupled dynamics of UAU–SIR model when compared to the MC simulations, especially for the cases of $\gamma > 0$.

In addition, we further check the role of infectivity reduction fact $\gamma$ in the coupling of disease spread and awareness diffusion in Figure 3.13. In panel (a), the infection and recovery rates of the disease spreading over the lower layer are fixed to be $\beta = 0.5$ and $\mu = 0.8$, while the loss rate of awareness is always set to be $\delta = 0.3$. If the awareness diffusion rate becomes higher (e.g., $\lambda = 0.8$), the stationary fraction of recovered individuals $\rho^R$ will be lower since the awareness can be spread into more individuals. Conversely, the lower diffusion rate (e.g., $\lambda = 0.2$) will lead to the larger $\rho^R$. However, whether $\lambda$ is higher or lower, larger $\gamma$ means that the risk reduction of aware individuals becomes weaker and our analytical results will be in good agreement with MC simulations, but $\gamma \to 0$ implies the infectivity reduction of aware ones will become much more obvious and the results predicted by the MMCA method will deviate a little from the MC simulations, which is also consistent with those in Figure 3.9. Similarly, in panel (b), we further examine the impact of $\gamma$ on the coupling spreading for different disease recovering rates in the disease spreading layer, and almost the identical qualitative behavior regarding the role of $\gamma$ can be observed here. Thus, $\gamma$ plays

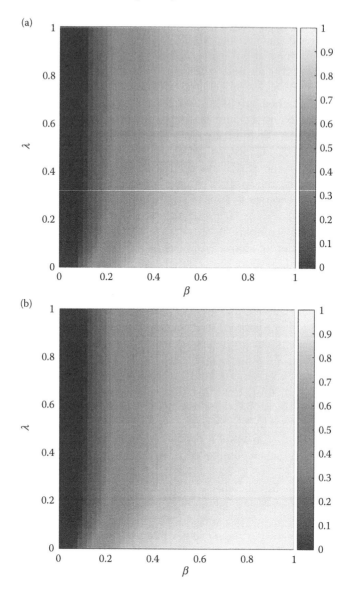

**FIGURE 3.11:** Comparisons between MMCA analytical results and MC simulations for the whole range of $\lambda$ and $\beta$ in our UAU–SIR model. The gray level values denote the fraction of recovered individuals ($\rho^R$) for each point within a grid of $50 \times 50$ at the stationary state, in which panel (a) represents the results of MMCA and panel (b) means those acquired by MC simulations averaged over 200 independent runs. The network setup is identical with those in Figures 3.9 and 3.10, the awareness loss rate is $\delta = 0.3$, the curing rate for infective nodes is $\mu = 0.8$ and the infectivity reduction factor among aware nodes $\gamma = 0.5$.

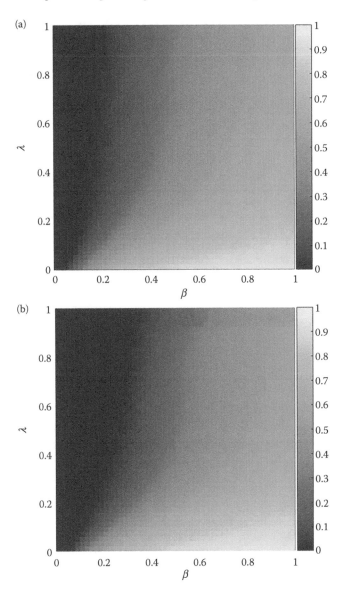

**FIGURE 3.12:** Comparisons between MMCA analytical results and MC simulations for the whole range of $\lambda$ and $\beta$ in our UAU–SIR model. The gray level values denote the fraction of recovered individuals ($\rho^R$) for each point within a grid of $50 \times 50$ at the stationary state, in which panel (a) represents the results of MMCA and panel (b) means those acquired by MC simulations averaged over 200 independent runs. The network setup is identical with those in Figures 3.9 and 3.10, the awareness loss rate is $\delta = 0.3$, the curing rate for infective nodes is $\mu = 0.8$ and the infectivity reduction factor among aware nodes $\gamma = 0.0$.

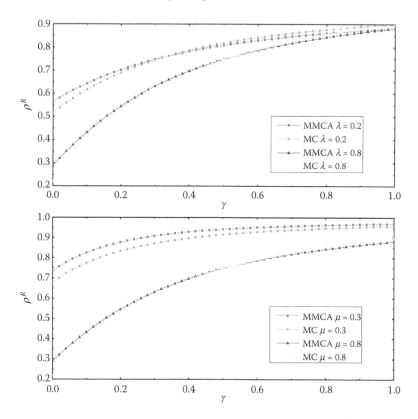

**FIGURE 3.13:** Stationary fraction of recovered individuals $\rho^R$ as a function of $\gamma$ for different values of $\lambda$, $\beta$, and $\delta$. The network setup is same as those in Figures 3.9 and 3.10, other parameters are set to be $\delta = 0.3$, $\beta = 0.5$, and $\mu = 0.8$ (a) and $\lambda = 0.8$, $\delta = 0.3$, and $\beta = 0.5$ (b).

an important role in characterizing the coupled spreading between disease and awareness, and we set $\gamma = 0.5$ in most simulations without loss of generality.

Finally, for the purpose of further cross-checking the analytical predictions in Equation 3.28, we depict the critical threshold of UAU–SIR model for different values of $\gamma$, $\mu$, and $\delta$ in the multiplex networks in Figure 3.14, and all these results indicate that MMCA is an effective prediction method for the coupled dynamics between the disease spreading and awareness diffusion. From panel (a) through (d), the simulation results indicate that the predicted critical threshold will become higher as the information spreading rate $\lambda$ increases, that is, the diffusion of information related with the epidemics will help to reduce the disease prevalence and restrain the epidemic spreading, which also conforms to the intuition in the practice of public health prevention. Additionally, these results well accord with those in Figure 3.9.

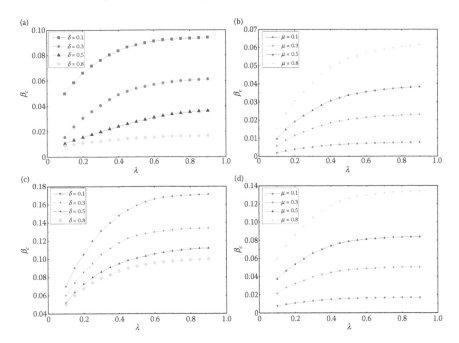

**FIGURE 3.14:** Epidemic threshold $\beta_c$ as a function of the information spreading rate $\lambda$ for different values of $\gamma$, $\mu$, and $\delta$. The network setup is identical with those in Figures 3.9 and 3.10, other parameters are set to be $\gamma = 0.5$ and $\mu = 0.8$ (a), $\gamma = 0.5$ and $\delta = 0.3$ (b), $\gamma = 0.0$ and $\mu = 0.8$ (c), and $\gamma = 0.5$ and $\delta = 0.3$ (d).

## 3.4 Concluding Remarks and Outlooks

In this chapter, we proposed two novel epidemic models in complex networks to assist in understanding the disease spreading behavior within the real-life systems. On the one hand, from the perspective of the single-layer networks, we consider the impact of nonuniform transmission and recovering delay on the epidemic spreading behavior by the mean-field theory analysis and Monte Carlo simulations, and find that the recovery delay will greatly alter the critical threshold for epidemic spreading in complex networks. On the other hand, according to the view of the multi-layer networks, we integrated the disease spreading and the awareness diffusion related with epidemics into one coupled disease spreading model, where the bottom layer is used to depict the epidemic spreading and the upper layer is used to characterize the corresponding awareness or information spreading in complex networks. Based on the MMCA, it is indicated that the prediction for the epidemic threshold well coincides with those results from MC numerical simulations. The current results are vastly useful to explore the role of disease infection scheme and topological structure in the epidemic spreading among the real population.

In the future, it becomes a much more relevant topic if we can further incorporate the individual behavior response into the mathematical modeling during the epidemic outbreak. Firstly, human preventive behavior may greatly reduce the risk of being infected from the individual perspective; secondly, the individual response may cut off some of links or cut down some contacts with other ones, even changing the topology of disease transmission; thirdly, collective social distancing sometimes loses the power at the expense of huge social costs and may be resisted by some agents for various factors, in particular, for the lower risk contagion. To summarize, how to enhance the accuracy or precision of modeling forecast is very crucial to perform the epidemic model when confronting various emerging epidemics.

## Acknowledgments

This project is partially supported by the Natural Science Foundation of China (NSFC) under grant nos. 61773286 and 61374169.

## References

1. H.W. Hethcote. The mathematics of infectious diseases. *SIAM Rev.* 42(4), 2000, 599–653.

2. D.M. Morens, G.K. Folkers, A.S. Fauci. The challenge of emerging and re-emerging infectious diseases. *Nature* 430(6996), 2004, 242–249.

3. W.T. Enanoria, L. Worden, F. Liu, D. Gao, S. Ackley, J. Scott, M. Deiner et al. Evaluating subcriticality during the ebola epidemic in west Africa. *PLoS ONE* 10(10), 2015, e0140651.

4. M. Lipsitch, T. Cohen, B. Cooper, J.M. Robins, S. Ma, L. James, G. Gopalakrishna et al. Transmission dynamics and control of severe acute respiratory syndrome. *Science* 300(5627), 2003, 1966–1970.

5. P.V. Mieghem, J. Omic, R. Kooij. Virus spread in networks. *IEEE ACM Trans. Netw.* 17(1), 2009, 1–14.

6. Y. Wang, S. Wen, Y. Xiang, W.L. Zhou. Modelling the propagation of worms in networks: A Survey. *IEEE Comm. Surv. Tutorials* 16(2), 2014, 942–960.

7. S. Wen, W. Zhou, J. Zhang, Y. Xiang, W.L. Zhou, W.J. Jia. Modeling propagation dynamics of social network worms. *IEEE Trans. Parallel Distrib. Syst.* 24(8), 2013, 1633–1643.

8. D. Henderson, P. Klepac. Lessons from the eradication of smallpox: An interview with DA Henderson, *Phil. Trans. R. Soc. B* 368(1623), 2013, 20130113.

9.  B. Bramanti, N.C. Stenseth, L. Wallфe, X. Lei. Plague: A disease which changed the path of human civilization. *Adv. Exp. Med. Biol.* 918, 2016, 1–26.

10. S. Eubank, H. Guclu, V.S.A. Kumar et al. Modelling disease outbreaks in realistic urban social networks. *Nature* 429(6988), 2004, 180–184.

11. Z. Wang, C.T. Bauchc, S. Bhattacharyya, A. d'Onofrio, P. Manfred, M. Perc, N. Perra, M. Salathé, D.W. Zhao. Statistical physics of vaccination. *Phys. Rep.* 664, 2016, 1–113.

12. M.J. Keeling, P. Rohani. *Modeling Infectious Diseases in Humans and Animals.* Princeton, NJ: Princeton University Press, 2008.

13. R.M. Anderson, R.M. Robert. *Infectious Diseases of Humans: Dynamics and Control.* New York, NY: Oxford University Press, 1992.

14. D. Bernoulli. Essai d'une nouvelle analyse de la mortalite causee par la pette verole. *Mem. Math. Phys. Acad. R. Sci.* Paris, 1766, 1–45.

15. W.O. Kermack, A.G. McKendrick. Contributions to the mathematical theory of epidemics. *Proc. R. Soc. A* 115, 1927, 700–721.

16. D.J. Watts, S.H. Strogatz. Collective dynamics of small-world networks. *Nature* 393 (6687), 1998, 440–442.

17. A.L. Barabási, R. Albert. Emergence of scaling in random networks. *Science* 286 (5439), 1999, 509–512.

18. R. Albert, A.L. Barabási. Statistical mechanics of complex networks. *Rev. Mod. Phys.* 74(1), 2002, 47–97.

19. S. Boccaletti, V. Latora, Y. Moreno, M. Chavezf, D.U. Hwang. Complex networks: Structure and dynamics. *Phys. Rep.* 424, 2006, 175–308.

20. R. Pastor-Satorras, C. Castellano, P.V. Mieghem, A. Vespignani. Epidemic processes in complex networks. *Rev. Mod. Phys.* 87, 2015, 925–979.

21. R. Pastor-Satorras, A. Vespignani. Epidemic spreading in scale-free networks. *Phys. Rev. Lett.* 86, 2001, 3200–3203.

22. V.M. Eguíluz, K. Klemm. Epidemic threshold in structured scale-free networks. *Phys. Rev. Lett.* 89, 2002, 108701.

23. G. Yan, T. Zhou, J. Wang et al. Epidemic spread in weighted scale-free networks. *Chin. Phys. Lett.* 22(2), 2005, 510–513.

24. Z.H. Liu, B. Hu. Epidemic spreading in community networks. *Europhys. Lett.* 72(2), 2005, 315–321.

25. W. Huang, C.G. Li. Epidemic spreading in scale-free networks with community structure. *J. Stat. Mech.* (1), 2008, P01014.

26. D.F. Zheng, P.M. Hu, S. Trimper. Epidemics and dimensionality in hierarchical networks. *Phys. A* 352, 2005, 659–668.

27. X.J. Xu, X. Zhang, J.F.F. Mendes. Impacts of preference and geography on epidemic spreading. *Phys. Rev. E* 76, 2007, 056109.

28. J. Joo, J.L. Lebowitz. Behavior of susceptible-infected-susceptible epidemics on heterogeneous networks with saturation. *Phys. Rev. E* 69, 2004, 066105.

29. R. Olinky, L. Stone. Unexpected epidemic threshold in heterogeneous networks: The role of disease transmission. *Phys. Rev. E* 70, 2004, 030902.

30. R. Yang, B.H. Wang, J. Ren et al. Epidemic spreading on heterogeneous networks with identical infectivity. *Phys. Lett. A* 364, 2007, 189–193.

31. J.Z. Wang, Z.R. Liu, J.H. Xu. Epidemic spreading on uncorrelated heterogeneous networks with non-uniform transmission. *Phys. A* 382, 2007, 715–721.

32. X.C. Fu, M. Small, D.M. Walker et al. Epidemic dynamics on scale-free networks with piecewise linear infectivity and immunization. *Phys. Rev. E* 77(3), 2008, 036113.

33. S. Boccaletti, G. Bianconi, R. Criado, C. Genio, J. Gómez-Gardenes, M. Romance, I. Sendina-Nadal, Z. Wang, M. Zanin. The structure and dynamics of multilayer networks. *Phys. Rep.* 544, 2014, 1–122.

34. C. Granell, S. Gómez, A. Arenas. Dynamical interplay between awareness and epidemic spreading in multiplex networks. *Phys. Rev. Lett.* 111(12), 2013, 128701.

35. C. Granell, S. Gómez, A. Arenas. Competing spreading processes on multiplex networks: Awareness and epidemics. *Phys. Rev. E* 90, 2014, 012808.

36. Q.T. Guo, Y.J. Lei, C.Y. Xia, L. Guo, X. Jiang, Z.M. Zheng. The role of node heterogeneity in the coupled spreading of epidemics and awareness. *PLoS ONE* 11(8), 2016, e0161037.

37. C.Y. Xia, Z. Wang, J. Sanz, S. Meloni, Y. Moreno. Effects of delayed recovery and nonuniform transmission on the spreading of diseases in complex networks. *Phys. A* 392(7), 2013, 1577–1585.

38. J.Q. Kan, H.F. Zhang. Effects of awareness diffusion and self-initiated awareness behavior on epidemic spreading-an approach on mulitplex networks. *Commun. Nonlinear Sci. Numer. Simul.* 44, 2017, 193–203.

39. S. Gómez, A. Arenas, J. Borge-Holthoefer, S. Meloni, Y. Moreno. Discrete-time Markov chain approach to contact-based disease spreading in complex networkks. *Europhys. Lett.* 89, 2010, 38009.

40. Y. Moreno, R. Pastor-Satorras, A. Vespignani. Epidemic outbreaks in complex heterogeneous networks. *Eur. Phys. J. B* 26(4), 2002, 521–529.

# Chapter 4

# Measurements for Investigating Complex Networks

Zengqiang Chen, Matthias Dehmer, and Yongtang Shi

## CONTENTS

## 4.1    Introduction

Complex networks have been utilized for establishing various models and practical applications, see References 69, 161, and 176. The definition of a complex network stems from the fact and observation that it should have a distinct topology. By definition, a complex network does not consist of regular and random graph patterns [69] and, hence, it only contains patterns possessing a distinct and nontrivial topology. As known, models based on complex networks have been applied interdisplinarly [69,161,176]; one plausible reason is their methodical core as it lies between graph theory and statistical mechanics. The analysis, discrimination, and synthesis of complex networks rely on the use of measurements for capturing the most relevant topological features [37,76]. Many kinds of measurements have been introduced to characterize the structure and properties of complex networks, including distance-based measurements, clustering coeffcient, degree correlation, graph entropies, centrality, subgraphs, spectral analysis, community-based measurements, hierarchical measurements, and fractal dimensions. In particular, Newman [161] reviewed a variety of techniques and models to improve the understanding of complex networks-based systems, including well-known graph classes like small world, scale free, and so forth. Another review is due to Costa et al. [37] which is a survey of measurements in the context of complex networks. We emphasize that graph theory plays an important role in the study of networks as it can be considered as its methodical core. Graph theory can be divided into two major categories: Classical Graph Theory, which is basically of descriptive nature [101] and Quantitative Graph Theory [44,49]. The latter deals with quantifying structural information of networks by using measurements, which belongs to a new branch of graph theory and network science. The main contribution of this chapter is to provide another review of complex networks complementing existing ones; we use the review due to Costa et al. [37]. The current review complements existing ones as it discusses measures which have not yet been discussed in survey papers. Also, the review surveys graph measures for random graphs (nondeterministic networks) and usual graphs (deterministic networks) leading to a compelling overview of measurements which can be used interdisciplinarily.

## 4.2    Distance-Based Measurements

The graphs we consider in this paper are simple and connected. Let $G$ be a graph with $n$ vertices and $m$ edges whose vertex set is $V(G)$ and edge set is $E(G)$. The distance between vertices $u$ and $v$ of $G$ is denoted by $d(u, v)$. The *eccentricity* of $v$ is $\sigma(v) = \max_{u \in V} d(u,v)$, where $d(u, v)$ is the distance between vertices $u$ and $v$. The *center* of $G$ is the set of vertices with minimum eccentricity. As with the centroid, the center of a tree consists of a single vertex or two adjacent vertices. The diameter of a graph $G$ has been defined by $D = D(G) = \max_{v \in V(G)} \sigma(v)$, see Reference 101.

The *Wiener index* of $G$ is denoted by $W(G)$ and is defined by

$$W(G) = \sum_{\{u,v\} \subseteq V(G)} d(u,v).$$

Consequently, the *average distance* between the vertices of $G$, denoted by $\mu(G)$, is given by

$$\mu(G) = W(G) / \binom{n(G)}{2}.$$

The name Wiener index or Wiener number for the quantity defined has been used in the chemical literature; Wiener [185] in 1947 seems to be the first who considered the measure. For more results on the Wiener index of trees, we refer to the survey [67]. In the mathematical literature, $W$ seems to be firstly studied in 1976 [78]; for a long time, mathematicians were unaware of the (earlier) work on $W$ done in chemistry. In the same paper, Wiener also introduced another index for acyclic molecules namely the *Wiener polarity index*. The Wiener polarity index of a graph $G = (V, E)$, denoted by $W_p(G)$, is the number of unordered pairs of vertices $\{u, v\}$ of $G$ such that $d_G(u, v) = 3$, i.e.,

$$W_p(G) := |\{\{u,v\} \mid d(u,v) = 3, u,v \in V(G)\}|.$$

Wiener used a liner formula of $W$ and $W_P$ to calculate the boiling points $t_B$ of the paraffns, i.e., $t_B = aW + bW_p + c$, where $a$, $b$, and $c$ are constants for a given isomeric group. Until now, many kinds of Wiener indices have been studied.

Except the Wiener indices, well-known distance measures include the Balaban index [8,9], Wiener polarity index [71,149,185], the Szeged index [93], the Schultz index [68,168], the Gutman index [94], the Harary index [109,163], and so on. After these indices were proposed, researchers studied their properties and also relations with other indices. For more results on mathematical properties of the Wiener index, we refer to the surveys [67,121]; and for more results on distance measures, we refer to the survey [187].

The cyclomatic number of $G$, denoted by $c(G)$, is the minimum number of edges that must be removed from $G$ in order to transform it to an acyclic graph. It is known that $c(G) = m - n + 1$. The *Balaban index* of a connected graph $G = (V, E)$ has been defined by

$$J(G) = \frac{m}{c(G) + 1} \sum_{uv \in E} \frac{1}{\sqrt{D_G(u)D_G(v)}}.$$

It has been proposed by Balaban [8,9] in 1982, which is also called the *average distance-sum connectivity index* or *Balaban J index*. Furthermore, Balaban et al. [10] proposed the concept of the *Sum-Balaban index* of a connected graph $G$. It has been defined by

$$SJ(G) = \frac{m}{c(G) + 1} \sum_{uv \in E} \frac{1}{\sqrt{D_G(u) + D_G(v)}}.$$

For an edge $e = ij$, let $n_e(i)$ be the number of vertices of $G$ being closer to $i$ than to $j$. The *Szeged index* of $G$ has been defined by Gutman [93]

$$Sz(G) = \sum_{e=ij} n_e(i)n_e(j),$$

during his stay at the Attila Jozsef University in Szeged.

In 1989, triggered by the idea of characterizing the alkanes, Schultz [168] defined a new index MTI(G) that is degree and distance based. Gutman decomposed this index into two parts and called one of them *Schultz index* (of the first kind), which is defined by

$$S(G) = \sum_{\{u,v\} \in V(G)} (d(u) + d(v))d(u,v),$$

where $d(v)$ denotes the degree of $v$. The same invariant was independently and simultaneously introduced by Dobrynin and Kochetova [68].

Gutman [94] also introduced a new index,

$$Gut(G) = \sum_{\{u,v\} \in V(G)} d(u)d(v)d(u,v),$$

and called it the Schultz index of the second kind. Nowadays, this index is also known as the *Gutman index.*

In 1993, Plavšić et al. [163] and Ivanciuc et al. [109] independently introduced the Harary index, named in honor of Frank Harary on the occasion of his 70th birthday. Actually, the Harary index has been defined by

$$H(G) = \sum_{\{u,v\} \in V(G)} \frac{1}{d(u,v)}.$$

## 4.3    Degree-Based Measurements

The degree is an important characteristic of a vertex, which is the number of edges incident to the vertex. Based on the degree of the vertices, it is possible to derive many measurements for networks. An important property of many real-world networks is their power law degree distribution [11]. Xiao et al. [186] studied the clique-degree distribution of networks. For more results on degree distributions of networks, see References 25, 170, and 195.

Topological indices are molecular descriptors which play an important role in theoretical chemistry, especially in QSPR/QSAR research [64]. Among all topological indices, one of the most investigated is the so-called degree-based topological index, defined for a graph $G$ with $n$ vertices as

$$TI(G) = \sum_{1 \leq i \leq j \leq n-1} m_{ij}\phi_{ij}, \tag{4.1}$$

where $m_{ij}$ is the number of edges of $G$ joining a vertex of degree $i$ with a vertex of degree $j$ and $\{\phi_{ij}\}$ is a set of real numbers. The well-known distance measures include various of Randić indices [133,164], Zagreb indices [98], the Harmonic index [192], the Geometric-arithmetic index [180], the sum-connectivity index [194], Atom-bond-connectivity index [80], and so on.

It seems that the Randić index was the first degree-based topological index. In 1975, the chemist Randić [164] proposed a topological index $R$ ($R_{-1}$ and $R_{-\frac{1}{2}}$) under the name "branching index" or "connectivity index", suitable for measuring the extent of branching of the carbon-atom skeleton of saturated hydrocarbons. Until now, this index is also called "Randić index". In 1998, Bollobás and Erdös [14] generalized this index by replacing $-\frac{1}{2}$ with any real number $\alpha$, which is called the *general Randić index*. Triggered by this development, the Randić index and general Randić index soon became the most popular and most frequently employed structural descriptor, used in innumerable QSPR and QSAR studies. For a graph $G = (V, E)$, the *Randić index* $R(G)$ of $G$ has been defined by the sum of $(d(u)d(v))^{-1/2}$ over all edges $uv$ of $G$, i.e.,

$$R(G) = \sum_{uv \in E} \frac{1}{\sqrt{(d(u)d(v))}},$$

where $d(u)$ denotes the degree of a vertex $u$ of $G$. The *zeroth-order Randić index*, defined by Kier and Hall [119], is

$$^0R(G) = \sum_{u \in V(G)} \frac{1}{\sqrt{d(u)}}.$$

The degree power is one of the most important graph invariant and well-studied in graph theory. The sum of degree powers of $G$ is defined by

$$\sum_{v \in V(G)} d(v)^\alpha,$$

where $\alpha$ is an arbitrary real number. This is also called *zeroth-order general Randić index* [117,118,133,164]. Actually, the sum of degree powers also has many applications in information theory, social networks, network reliability, and mathematical chemistry. For more results on the Randić index and the zeroth-order Randić index, we refer to the two surveys of Li and Shi [133,135], and the book due to Li and Gutman [130].

Now we consider the expression (4.1). From the above, we know that Randić index is defined by $\phi_{ij} = \frac{1}{\sqrt{ij}}$. The first Zagreb index [98] is defined by $\phi_{ij} = i + j$, while the second Zagreb index [98] is defined by $\phi_{ij} = ij$. The Harmonic index [192] is defined by $\phi_{ij} = \frac{2}{i+j}$. The Geometric-arithmetic index [180] is defined by $\phi_{ij} = \frac{2\sqrt{ij}}{i+j}$. The sum-connectivity index [194] is defined by $\phi_{ij} = \frac{1}{\sqrt{i+j}}$. The Atom-bond-connectivity index [80] is defined by $\phi_{ij} = \sqrt{\frac{i+j-2}{ij}}$.

## 4.4   Spectral Measurements

Let $G = (V, E)$ be a graph with $n$ vertices and $m$ edges. Let $A(G)$ be the adjacency matrix of $G$. The eigenvalues $\lambda_1, \lambda_2, \ldots, \lambda_n$ of the the matrix $A(G)$ are said to be the eigenvalues of graph $G$ to form its spectrum. The $k$-th *spectral moment* of graph $G$ is defined by $M_k(G) = \sum_{i=1}^{n} \lambda_i^k$. Observe that for odd $k$, $M_k(G) = 0$ if $G$ is a bipartite graph. In order to overcome this limitation, the authors in Reference 193 defined the moment-like quantities, $M_k^*(G) = \sum_{i=1}^{n} |\lambda_i|^k$. Details of the spectral theory of graphs can be found in the seminal monograph [36].

One of the most remarkable chemical application of graph theory is based on the close correspondence between the graph eigenvalues and the molecular orbital energy levels of $\pi$-electrons in conjugated hydrocarbons. In the seventies, Gutman [92] introduced the following definition of graph energy:

**Definition 4.1.** *If $G$ is a graph on $n$ vertices and $\lambda_1, \lambda_2, \ldots, \lambda_n$ are its eigenvalues, then the energy of $G$ is $E(G) = \sum\limits_{i=1}^{n} |\lambda_i|$.*

In the theory of graph energy the so-called *Coulson integral formula* plays an important role. This formula was obtained by Charles Coulson in 1940 [34]:

$$\mathcal{E}(G) = \frac{1}{\pi} \int\limits_{-\infty}^{+\infty} \left[ n - \frac{ix\phi'(G,ix)}{\phi(G,ix)} \right] dx = \frac{1}{\pi} \int\limits_{-\infty}^{+\infty} \left[ n - x\frac{d}{dx}\ln\phi(G,ix) \right] dx,$$

where $G$ is a graph, $\phi(G, x)$ is the characteristic polynomial of $G$, $\phi'(G, x) = (d/dx)\phi(G, x)$ its first derivative, and $i = \sqrt{-1}$. For more details on this useful equality, we refer to References 34 and 134.

There have been two important classes of mathematical problems on graph energy. One class is to find the upper and lower bounds of graph energy. Another relates to determine the extremal values of the energy for a given class of graphs, and also characterize the corresponding extremal graphs. Some of these results are as follows. For a graph $G$ with $m$ edges, we have $2\sqrt{m} \leq \mathcal{E}(G) \leq 2m$ [152]. Let $G$ be a graph with $n$ vertices and $m$ edges. Then $\mathcal{E}(G) \leq \sqrt{2mn}$ [152]. Koolen and Moulton [123,126] obtained the following result: If $2m \geq n$, then

$$\mathcal{E}(G) \leq \frac{2m}{n} + \sqrt{(n-1)\left[ 2m - \left( \frac{2m}{n} \right)^2 \right]}.$$

If, in addition, $G$ is bipartite, then [125,126]

$$\mathcal{E}(G) \leq \frac{4m}{n} + \sqrt{(n-2)\left[ 2m - 2\left( \frac{2m}{n} \right)^2 \right]}.$$

Let $T$ be a tree of order $n$. A basic result is $\mathcal{E}(S_n) \leq \mathcal{E}(T) \leq \mathcal{E}(P_n)$, see Reference 91; $S_n$ and $P_n$ denote the star graph and path graph of order $n$, respectively. The unicyclic graphs with maximum energy are finally determined in References 7 and 106, independently. Huo et al. [105] determined the maximal energy among all bipartite bicyclic graphs. Wagner [181] showed that the maximum value of the graph energy within the set of all graphs with cyclomatic number $k$ (which includes, for instance, trees or unicyclic graphs as special cases) is at most $4n/\pi + c_k$ for some constant $c_k$ that only depends on $k$. We also mention some results due to Dehmer et al. [51,62,63]. These contributions put the emphasis on defining graph measures based on the eigenvalues of graph polynomials by using special graph-theoretical matrices. As a result, the measures turned out to be highly discriminating and even outperformed some of the well-known molecular identification numbers developed by Randić [165]. For more results on graph energy, see References 92, 97, and 134.

Nowadays, besides the adjacency matrix, many kinds of graph matrices and their spectrums are investigated, such as incidence matrix, (signless) Laplacian matrix, and distance matrix. Therefore, many kinds of graph energies are introduced and studied, including matching energy [99], energy of matrix [162], Laplacian energy [100], Randić energy [18], incidence energy [112], distance energy [108], skew energy [2], resolvent energy [95], and so on. For more details, see Reference 96.

Let $G$ be a simple, connected graph of order $n$ with eigenvalues $\lambda_1 \geq \lambda_2 \geq \cdots \geq \lambda_n$. Actually, $\lambda_1$ is called the spectral radius of $G$, which is widely studied. The *median eigenvalues* $\lambda_H$, $\lambda_L$, for $H = \lfloor (n+1)/2 \rfloor$ and $L = \lceil (n+1)/2 \rceil$ are considered. These play an important role in the Hückel molecular orbital model of $\pi$-electron systems. The *HL*-index is defined as $R(G) := \max\{|\lambda_H|, |\lambda_L|\}$ by Jaklič et al. [110]. For more results on median eigenvalues and HL-index, we refer to References 131, 155–158.

## 4.5   Network Similarity

Methods to determine the structural similarity or distance between networks have been utilized in many areas of sciences. For example, in mathematics [55,174,189], in biology [75,114,119], in chemistry [17,173], and in chemoinformatics [178]. Other application-oriented areas where graph comparison techniques have been employed can be found in References 55, 103, and 153. To measure the distance or similarity between distinct networks has been an interesting and useful task in applied mathematics and related disciplines. However, finding the right measure/method for practical problems has been intricate as structural distance or similarity is in the eye of a beholder. There are many survey papers on network similarities. Bunke [20] puts the emphasis on inexact graph matching methods applied to image and video indexing but, for example, does not discuss classical contributions dealing with graph isomorphism (exact graph matching). Conte et al. [32] discussed graph matching techniques for image, document, and video analysis and the presented taxonomy is solely focused on algorithms from these areas and also does not discuss the complexity of graph matching. Gao et al. [88] focused on surveying results for the graph edit distance only. Very recently, Emmert-Streib et al. [77] presented a survey on graph matching, network comparison, and network alignment methods, which are more comprehensive. For more results on network similarities, see References 140, 141, and 145.

The development of methods for the comparison of networks has been an active area of research and in recent years, many new methods have been introduced. We emphasize one of the first quantitative graph distance measure based on isomorphic relations which has been developed by Zelinka [189]. Also, Kaden [115] and Sobik [174] further developed Zelinka's approach. A measure from another paradigm namely the *graph edit distance* (GED) has been introduced by Bunke [20]. The principle idea of the GED is to define so-called *graph edit operations* such as insertions or deletions of edges/vertices or relabelings of vertices along with certain edit costs associated with the mentioned graph edit operations. The graph edit distance of two given graphs is the minimum cost associated with a series of edit operations.

Dehmer [42] examined relations between information-theoretic measures which are based on information functionals and between classical and parametric graph entropies [59]. Also Dehmer et al. put the emphasis on graph distance measures which are based on so-called topological indices, see Reference 48. Applications of graph similarity when dealing with link prediction in web mining are due to Lu and Zhou [146,147].

## 4.6    Network Entropy

The study of entropy measures for exploring network-based systems emerged in the late fifties based on the seminal work due to Shannon [169]. Then Rashevsky [166] and Mowshowitz [159] studied mathematical properties of the measures thoroughly and also discussed applications thereof. As known, graph entropy measures have been used in various disciplines, for example, for characterizing graph patterns in biology, chemistry, and computer science, see References 4, 26, 46, 128, and 182. Thus, it is not surprising to see that term 'graph entropy' has been defined in various ways; each discipline and application may require a different definition. An example is the Körner entropy [124] introduced from an information theory-specific point of view versus the graph entropy measures of Mowshowitz [159] which can be understood as indices for measuring symmetry in graphs.

Note that several graph invariants, such as the number of vertices, the vertex degree sequences, extended degree sequences (the second neighbor, third neighbor, etc.), eigenvalues, and connectivity information, have been used for developing graph entropy measures, see References 29, 42, 58, and 70. Distance-based graph entropies [29,42] are also studied, which is related to the average distance and various Wiener indices [3,38,83,84,120,138,139,150,172,175,185]. For more results on network entropy, see References 45 and 58. As reported by Dehmer and Kraus [53], there is a lack of analytical results when proving extremal results for entropy-based graph measures. Recently, some extremal properties of kinds of entropies by using some theoretical methods are investigated.

In order to start, we reproduce the definition of Shannon's entropy [169]. In the whole paper, "log" denotes the logarithm based on 2.

**Definition 4.2.** *Let* $p = (p_1, p_2, \ldots, p_n)$ *be a probability vector, namely,* $0 \leq p_i \leq 1$ *and* $\sum_{i=1}^{n} p_i = 1$. *The Shannon's entropy of p is defined as*

$$I(p) = -\sum_{i=1}^{n} p_i \log p_i.$$

To define information-theoretic graph measures, we will often consider a tuple $(\lambda_1, \lambda_2, \ldots, \lambda_n)$ of nonnegative integers $\lambda_i \in \mathbb{N}$ [42]. This tuple forms a probability distribution $p = (p_1, p_2, \ldots, p_n)$, where

$$p_i = \frac{\lambda_i}{\sum_{j=1}^{n} \lambda_j} \quad i = 1, 2, \ldots, n.$$

Therefore, the entropy of tuple $(\lambda_1, \lambda_2, \ldots, \lambda_n)$ is given by

$$I(\lambda_1, \lambda_2, \ldots, \lambda_n) = -\sum_{i=1}^{n} p_i \log p_i = \log\left(\sum_{i=1}^{n} \lambda_i\right) - \sum_{i=1}^{n} \frac{\lambda_i}{\sum_{j=1}^{n} \lambda_j} \log \lambda_i. \quad (4.2)$$

In the literature, there are various ways to obtain the tuple $(\lambda_1, \lambda_2, \ldots, \lambda_n)$, like the so-called magnitude-based information measures introduced by Bonchev and Trinajstić [17], or partition-independent graph entropies, introduced by Dehmer [42,65], which are based on information functionals.

We are now ready to define the entropy of a graph due to Dehmer [42] by using information functionals.

**Definition 4.3.** *Let* $G = (V, E)$ *be a connected graph. For a vertex* $v_i \in V$, *we define*

$$p(v_i) := \frac{f(v_i)}{\sum_{j=1}^{|V|} f(v_j)}, \quad (4.3)$$

*where f represents an arbitrary information functional.*

Observe that $\sum_{i=1}^{|V|} p(v_i) = 1$. Hence, we can interpret the quantities $p(v_i)$ as vertex probabilities.

Now we immediately obtain one definition of a graph entropy of graph $G$.

**Definition 4.4.** *Let $G = (V, E)$ be a connected graph and $f$ be an arbitrary information functional. The entropy of $G$ is defined as*

$$I_f(G) = -\sum_{i=1}^{|V|} \frac{f(v_i)}{\sum_{j=1}^{|V|} f(v_j)} \log\left(\frac{f(v_i)}{\sum_{j=1}^{|V|} f(v_j)}\right)$$

$$= \log\left(\sum_{i=1}^{|V|} f(v_i)\right) - \sum_{i=1}^{|V|} \frac{f(v_i)}{\sum_{j=1}^{|V|} f(v_j)} \log f(v_i). \tag{4.4}$$

For more results on network entropy, see References 45 and 58.

## 4.6.1    Distance-based entropy

Distance in graphs has been one of the most important graph invariants. We first restate some definitions of the information functionals [42] based on distances. In Reference 42, the following information functional was introduced:

$$f(v_i) = \alpha^{\sum_{j=1}^{D(G)} c_j |S_j(v_i, G)|},$$

where $c_j$ with $j = 1, 2, \dots, D(G)$ and $\alpha$ are arbitrary real positive parameters. The graph entropy measure proposed in Reference 122 is calculated for a vertex $v_i$ as the entropy of its shortest distances from all other vertices in the graph:

$$H(v_i) = -\sum_{u \in V} \frac{d(v_i, u)}{D(v_i)} \log \frac{d(v_i, u)}{D(v_i)},$$

where $D(v_i) = \sum_{u \in V} d(v_i, u)$. The aggregation function over all distances of vertices in the graph is proposed as follows:

$$H = \sum_{v \in V} H(v_i).$$

The information functional based on the shortest distances is introduced in Reference 1:

$$f(v_i) = \sum_{u \in V} d(v_i, u).$$

There are also some functionals based on the betweenness centralities [1,19]. For more results on distance-based entropies, we refer to References 27 and 29.

In Reference 41, Das and Shi studied the following entropy based on graph eccentricities and some mathematical properties are characterized.

Let $G = (V, E)$. For a vertex $v_i \in V$, we define $f$ as

$$f(v_i) := c_i \sigma(v_i),$$

where $c_i > 0$ and $\sigma(v_i) = \max\{d(v_i, v) : v \in V(G)\}$ for $1 \le i \le n$. From Equation 4.4, the entropy based on $f$, denoted by $If_\sigma(G)$, is defined as follows:

$$If_\sigma(G) = \log\left(\sum_{i=1}^n c_i\sigma(v_i)\right) - \sum_{i=1}^n \frac{c_i\sigma(v_i)}{\sum_{j=1}^n c_j\sigma(v_j)} \log(c_i\sigma(v_i)).$$

## 4.6.2 Number of $k$-paths

In this subsection, we consider a new information functional, which is the number of vertices with distance $k$ to a given vertex [29]. For a given vertex $v$ in a graph, the number of vertices with distance one to $v$ is exactly the degree of $v$. On the other hand, the number of pairs of vertices with distance three, which is also related to the clustering coeffcient of networks, is also called the Wiener polarity index introduced for molecular networks by Wiener in 1947 [185].

Let $G = (V, E)$ be a connected graph with $n$ vertices and $v_i \in V(G)$. Denoted by $n_k(v_i)$ the number of vertices with distance $k$ to $v_i$, i.e.,

$$n_k(v_i) = |S_k(v_i, G)| = |\{u : d(u, v_i) = k,\ u \in V(G)\}|,$$

where $k$ is an integer such that $1 \le k \le D(G)$.

**Definition 4.5.** *Let $G = (V, E)$ be a connected graph. For a vertex $v_i \in V$ and $1 \le k \le D(G)$, we define the information functional as:*

$$f(v_i) := n_k(v_i).$$

Therefore, by applying Definition 4.5 and Equality (4.2), we obtain the special graph entropy

$$I_k(G) := I_f(G) = -\sum_{i=1}^n \frac{n_k(v_i)}{\sum_{j=1}^n n_k(v_j)} \log\left(\frac{n_k(v_i)}{\sum_{j=1}^n n_k(v_j)}\right)$$

$$= \log\left(\sum_{i=1}^n n_k(v_i)\right) - \frac{1}{\sum_{j=1}^n n_k(v_j)} \cdot \sum_{i=1}^n n_k(v_i) \log n_k(v_i). \qquad (4.5)$$

In Reference 29, Chen et al. discuss the extremal properties of the above graph entropy.

## 4.6.3 Eigenvalue-based entropy

Dehmer et al. also developed a graph entropy measure by using the moduli of the eigenvalues of a graph by employing several graph theoretical matrices [63].

They proved that this measure (among others) has high discrimination power by using chemical structures and exhaustively generated graphs.

In Reference 57, Dehmer and Mowshowitz introduced a new class of measures (called here generalized measures) that derive from functions such as those defined by Rényi's entropy [167] and Daròczy's entropy [39].

**Definition 4.6.** *Let $G$ be a graph of order $n$. Then*

i. $\displaystyle I^1(G) := \sum_{i=1}^{n} \frac{f(v_i)}{\sum_{j=1}^{n} f(v_j)} \left[ 1 - \frac{f(v_i)}{\sum_{j=1}^{n} f(v_j)} \right],$

ii. $\displaystyle I_\alpha^2(G) := \frac{1}{1-\alpha} \log \left( \sum_{i=1}^{n} \left( \frac{f(v_i)}{\sum_{j=1}^{n} f(v_j)} \right)^\alpha \right), \quad \alpha \neq 1,$

iii.
$$I_\alpha^3(G) := \frac{\sum_{i=1}^{n} \left( \dfrac{f(v_i)}{\sum_{j=1}^{n} f(v_j)} \right)^\alpha - 1}{2^{1-\alpha} - 1}, \quad \alpha \neq 1.$$

Let $G$ be an undirected graph of order $n$ and $A$ its adjacency matrix. Denote by $\lambda_1, \lambda_2, \ldots, \lambda_n$, the eigenvalues of $G$. If $f := |\lambda_i|$, then

$$p^f(v_i) = \frac{|\lambda_i|}{\sum_{j=1}^{n} |\lambda_i|}.$$

Therefore, the generalized graph entropies are as follows:

i. $\displaystyle I^1(G) := \sum_{i=1}^{n} \frac{|\lambda_i|}{\sum_{j=1}^{n} |\lambda_i|} \left[ 1 - \frac{|\lambda_i|}{\sum_{j=1}^{n} |\lambda_i|} \right],$ \hfill (4.6)

ii. $\displaystyle I_\alpha^2(G) := \frac{1}{1-\alpha} \log \left( \sum_{i=1}^{n} \left( \frac{|\lambda_i|}{\sum_{j=1}^{n} |\lambda_i|} \right)^\alpha \right), \quad \alpha \neq 1,$ \hfill (4.7)

iii.
$$I_\alpha^3(G) := \frac{\sum_{i=1}^{n} \left( \dfrac{|\lambda_i|}{\sum_{j=1}^{n} |\lambda_i|} \right)^\alpha - 1}{2^{1-\alpha} - 1}, \quad \alpha \neq 1. \tag{4.8}$$

In Reference 54, Dehmer et al. examine the extremal values of the above-stated entropies in terms of graph energy and the spectral moments. Using the similar method, some extremal properties of the generalized graph entropies by employing graph energies and other topological indices are investigated in Reference 132.

### 4.6.4 Degree based entropy

In Reference 23, Cao et al. studied novel properties of graph entropies which are based on a special information functional by using degree powers of graphs.

**Definition 4.7.** *Let $G = (V, E)$ be a connected graph. For a vertex $v_i \in V$, we define the information functional as:*

$$f := d_i^k,$$

*where $d_i$ is the degree of vertex $v_i$ and $k$ is an arbitrary real number.*

Therefore, by applying Definition 4.7 and Equality (4.4), we obtain the special graph entropy

$$I_f(G) = -\sum_{i=1}^{n} \frac{d_i^k}{\sum_{j=1}^{n} d_j^k} \log\left(\frac{d_i^k}{\sum_{j=1}^{n} d_j^k}\right)$$

$$= \log\left(\sum_{i=1}^{n} d_i^k\right) - \sum_{i=1}^{n} \frac{d_i^k}{\sum_{j=1}^{n} d_j^k} \log d_i^k. \tag{4.9}$$

After it was introduced, the entropy has been widely studied and many related papers on theoretical aspects [22,30,40,41,107,143,144] and application aspects [111,154,179,197] are published. Very recently, Hu et al. [104] extend the results of graph entropy based on degrees to uniform hypergraphs.

### 4.6.5 Entropy for weighted networks

In order to investigate the influence of the structure of social relations between individuals of community's economic development, Eagle et al. [73] developed two new metrics, social diversity and spatial diversity, to capture the social and spatial diversity of communication ties within a social network of each individual, by using the entropy for vertices. Following this, in Reference 28, we introduce the concept of graph entropy for weighted graphs. We mention that Dehmer et al. [43] already tackled the problem of defining the entropy of weighted chemical graphs by using special information functionals. So, in Reference 28, we extend the work done in Reference 43 considerably.

Here we use the class of weighted graphs due to Bollobás and Erdős, which is called the general Randić index. Let $I(G,\alpha)$ be the entropy $I(G, w)$ based on the above-stated weight, i.e.,

$$I(G,\alpha) = -\sum_{uv \in E} \frac{(d(u)d(v))^\alpha}{\sum_{uv \in E}(d(u)d(v))^\alpha} \log\left(\frac{(d(u)d(v))^\alpha}{\sum_{uv \in E}(d(u)d(v))^\alpha}\right).$$

The above equality can also be expressed as

$$I(G,\alpha) = \log(R_\alpha(G)) - \frac{\alpha}{R_\alpha(G)} \sum_{uv \in E} (d(u)d(v))^\alpha \log(d(u)d(v)).$$

In Reference 28, the extremal values and examine extremal properties of this entropy are studied.

---

## 4.7   Other Measurements

### 4.7.1   Vulnerability and reliability measurements

There are several measures or graph parameters of the vulnerability and reliability of a communication network, such as connectivity, toughness, scattering number, integrity, tenacity, rupture degree, and edge-analogues of some of them. Vertex connectivity and edge connectivity are well studied in graph theory and networks, which can be found in any textbook of graph theory.

Network vulnerability was studied by the researchers [90,129]. The *vulnerability* of a vertex can be defined as the drop in performance when the vertex and all its edges are removed from the network

$$V_i = \frac{E - E_i}{E}$$

where $E$ is the global effciency of the original network and $E_i$ is the global effciency after the removal of the vertex $i$ and all its edges.

The concept of scattering number was first introduced by Jung [113]. For a graph $G = (V, E)$, the *scattering number* of $G$ is defined as

$$s(G) = \max\{\omega(G - X) - |X| : X \subseteq V(G), \omega(G - X) \neq 1\}.$$

Jung pointed out that the scattering number is the minimal number of disjoint paths that cover the vertices of graph $G$ and $s(G)$ is in a certain sense the "additive dual" for the concept of toughness, which was defined by Chvátal [31] as $t(G) = \min\{\frac{|X|}{\omega(G-X)} : X \subseteq V(G), \omega(G - X) \neq 1\}$.

To disrupt the network that is modeled as a graph, a terrorist attempts to remove a small set of vertices such that the remaining connected components are small. The *integrity of G* is defined as

$$I(G) = \min_{X \subset V}\{|X| + m(G - X)\},$$

which was introduced by Barefoot et al. [12].

Another reasonable measure of network vulnerability and reliability is called the tenacity of graphs, which was introduced by Cozzens et al. [35]. The *tenacity* $T(G)$ of $G$ is defined as

$$T(G) = \min\left\{\frac{|X| + m(G - X)}{\omega(G - X)}\right\}.$$

Many properties of these measurements are studied. For example, in Reference 137, Li et al. studied the spectrum bounds for the scattering number, integrity, and tenacity of regular graphs.

## 4.7.2 Subnetwork structural measurements

The structural properties of subnetworks are investigated to characterize the structures of networks. A characteristic of the Erdös–Rényi model is that the local structure of the network near a vertex tends to be a tree. Some measurements proposed to study the cyclic structure of networks and the tendency to form sets of tightly connected vertices are described.

The concept of clustering coefficient is frequently used. Let $N_\Delta$ be the number of triangles in the network and $N_3$ the number of connected triples. For an undirected unweighted network $G$, the clustering coefficient is defined as follows [160]:

$$C(G) = \frac{3N_\Delta}{N_3}.$$

Note that the clustering coefficient of a network is related to the Wiener polarity index which is the number of pairs of vertices with distance three, introduced in Section 4.2. Actually, Bollobás et al. studied the number of paths with length $k$ in graphs with $n$ vertices [15,16], while Erdös and Alon studied the number of given subgraphs in graphs with $n$ vertices [5,6,79].

## 4.7.3 Community structure

Large-scale networks with thousand to million of nodes are ubiquitous across many different scientific domains. Detecting community structures in these networks is often of particular interest [102]. Identifying communities in a network is a complex problem due to the existence of numerous definitions of community and the intractability of many community detection algorithms. In recent years, several surveys in the area of community detection have been published [33,85, 102,136,151].

## 4.7.4 Polynomial-based measurements

Graph polynomials have been proved useful in several scientific disciplines such as discrete mathematics, engineering, information sciences, mathematical chemistry, and related disciplines.

In general, a graph polynomial can encode structural information about the underlying graph in various ways. Examples are the characteristic or chromatic polynomial of a graph. In addition, graph polynomials and their zeros have been a valuable source for investigating various problems in discrete mathematics and related areas. The very first polynomial investigated in graph theory has been introduced by Sylvester in 1878, and was further studied by Petersen [198]. Until now, plenty of graph polynomials have been developed, such as the matching polynomial [82,89], permanental polynomial [142], Hosoya polynomial [66], Tutte polynomial [74,177], and so forth. For more results on graph polynomials, we refer to Reference 171.

Some network measurements based on the roots of various of graph polynomials have been introduced by Dehmer et al. [47,50,52,56,60–62]. One of the important future work on this topic is to explore properties of quantitative network measures which are based on graph polynomials.

### 4.7.5    Centrality measurements

In networks, the greater the number of paths in which a vertex or edge participates, the higher the importance of this vertex or edge for the network. The *betweenness centrality* [86] is defined as:

$$B_u = \sum_{ij} \frac{\sigma(i,u,j)}{\sigma(i,j)},$$

where $\sigma(i, u, j)$ is the number of shortest paths between vertices $i$ and $j$ that pass through vertex or edge $u$, $\sigma(i, j)$ is the total number of shortest paths between $i$ and $j$, and the sum is over all pairs $i$, $j$ of distinct vertices. Other centrality measurements can be found in the interesting survey by Koschützki et al. [127].

### 4.7.6    Synchronization phenomenon

Synchronization phenomenon in networks is also well studied, e.g., Wang and Chen [183]. Chen and Duan [24] address the fundamental problem of complex network synchronizability from a graph theoretic approach. Some parameters are used to estimate network synchronizability: betweenness [184], average distance [191], community structures [196], and substructures [72]. For more results, see References 21, 188, and 190.

---

## 4.8    Concluding Remarks

Measurements of the connectivity and topology of complex networks are essential for the characterization, analysis, classification, modeling, and validation of complex networks. The current survey has been organized to provide a comprehensive coverage from more graphical and mathematical view. Nowadays, except the general undirected networks, many other kinds of networks are widely studied, such as directed networks [148], weighted networks, networks of networks [13,81,87,116], and so on. Based on this, more measurements or invariants will be introduced for the further study of properties of different kinds of networks.

# References

1. O. Abramov and T. Lokot. Typology by means of language networks: Applying information theoretic measures to morphological derivation networks. In M. Dehmer, F. Emmert-Streib, and A. Mehler, editors, *Towards an Information Theory of Complex Networks: Statistical Methods and Applications*, pages 321–346. Springer, Boston, US, 2011.

2. C. Adiga, R. Balakrishnan, and W. So. The skew energy of a digraph. *Linear Algebra Appl.*, 432:1825–1835, 2010.

3. T. Al-Fozan, P. Manuel, I. Rajasingh, and R.S. Rajan. Computing Szeged index of certain nanosheets using partition technique. *MATCH Commun. Math. Comput. Chem.*, 72:339–353, 2014.

4. E.B. Allen. Measuring graph abstractions of software: An information-theory approach. In *Proceedings of the 8th International Symposium on Software Metrics Table of Contents*, page 182. IEEE Computer Society, Ottawa, Canada, 2002.

5. N. Alon. On the number of subgraphs of prescribed type of graphs with a given number of edges. *Isr. J. Math.*, 38:116–130, 1981.

6. N. Alon. On the number of certain subgraphs contained in graphs with a given number of edges. *Isr. J. Math.*, 53:97–120, 1986.

7. E.O.D. Andriantiana and S. Wagner. Unicyclic graphs with large energy. *Linear Algebra Appl.*, 435:1399–1414, 2011.

8. A.T. Balaban. Highly discriming distance-based topological index. *Chem. Phys. Lett.*, 89:399–404, 1982.

9. A.T. Balaban. Topological indices based on topological distance in molecular graphs. *Pure Appl. Chem.*, 55:199–206, 1983.

10. A.T. Balaban, P.V. Khadikar, and S. Aziz. Comparison of topological indices based on iterated 'sum' versus 'product' operations. *Iran. J. Math. Chem.*, 1:43–67, 2010.

11. A.-L. Barabási and R. Albert. Emergence of scaling in random networks. *Science*, 286:509–512, 1999.

12. C.A. Barefoot, R. Entringer, and H. Swart. Vulnerability in graphs—A comparative survey. *J. Combin. Math. Combin. Comput.*, 1:12–22, 1987.

13. S. Boccaletti, G. Bianconi, R. Criado, C.I. del Genio, J. Gómez-Garde nesi, M. Romance, I. Sendi na Nadal, Z. Wang, and M. Zaninmn. The structure and dynamics of layer networks. *Phys. Rep.*, 544(1):1–122, 2014.

14. B. Bollobás and P. Erdös. Graphs of extremal weights. *Ars Combin.*, 50:225–233, 1998.

15. B. Bollobás and A. Sarkar. Paths in graphs. *Stud. Sci. Math. Hung.*, 38:115–137, 2001.

16. B. Bollobás and M. Tyomkyn. Walks and paths in trees. *J. Graph Theory*, 70 (1):54–66, 2012.

17. D. Bonchev and N. Trinajstić. Information theory, distance matrix and molecular branching. *J. Chem. Phys.*, 67:4517–4533, 1977.

18. S.B. Bozkurt, A.D. Güngör, I. Gutman, and A.S. Cevik. Randić matrix and Randić energy. *MATCH Commun. Math. Comput. Chem.*, 64:239–250, 2010.

19. U. Brandes. A faster algorithm for betweenness centrality. *J. Math. Sociol.*, 25:163–177, 2011.

20. H. Bunke. Recent developments in graph matching. In *15th International Conference on Pattern Recognition*, Barcelona, Spain, pages 117–124, 2000.

21. J. Cao and J. Lu. Adaptive synchronization of neural networks with or without time-varying delay. *Chaos*, 16(1): 013133, 2006.

22. S. Cao and M. Dehmer. Degree-based entropies of networks revisited. *Appl. Math. Comput.*, 261:141–147, 2015.

23. S. Cao, M. Dehmer, and Y. Shi. Extremality of degree-based graph entropies. *Inform. Sci.*, 278:22–33, 2014.

24. G. Chen and Z. Duan. Network synchronizability analysis: A graph-theoretic approach. *Chaos*, 18:037102, 2008.

25. Q. Chen and D. Shi. Markov chains theory for scale-free network. *Phys. A*, 360:121–133, 2006.

26. Y. Chen, K. Wu, X. Chen, C. Tang, and Q. Zhu. An entropy-based uncertainty measurement approach in neighborhood systems. *Inform. Sci.*, 279:239–250, 2014.

27. Z. Chen, M. Dehmer, F. Emmert-Streib, and Y. Shi. Entropy bounds for dendrimers. *Appl. Math. Comput.*, 242:462–472, 2014.

28. Z. Chen, M. Dehmer, F. Emmert-Streib, and Y. Shi. Entropy of weighted graphs with Randić weights. *Entropy*, 17:3710–3723, 2015.

29. Z. Chen, M. Dehmer, and Y. Shi. A note on distance-based graph entropies. *Entropy*, 16(10):5416–5427, 2014.

30. Z. Chen, M. Dehmer, and Y. Shi. Bounds for degree-based network entropies. *Appl. Math. Comput.*, 265:983–993, 2015.

31. V. Chvátal. Tough graphs and hamiltonian circuits. *Discrete Math.*, 306:910–917, 2006.

32. D. Conte, F. Foggia, C. Sansone, and M. Vento. Thirty years of graph matching in pattern regocnition. *Int. J. Pattern Recognit. Artif. Intell.*, 18:265–298, 2004.

33. M. Coscia, F. Giannotti, and D. Pedreschi. A classification for community discovery methods in complex networks. *Stat. Anal. Data Mining*, 4:512–546, 2011.

34. C.A. Coulson. On the calculation of the energy in unsaturated hydrocarbon molecules. *Proc. Camb. Phil. Soc.*, 36(2):201–203, 1940.

35. M. Cozzens, D. Moazzami, and S. Stueckle. The tenacity of a graph. In *Proceedings of Seventh International Conference on the Theory and Applications of Graphs*. Wiley, New York, NY, 1995.

36. D. Cvetković, M. Doob, and H. Sachs. *Spectra of Graphs—Theory and Application*. Academic Press, New York, NY, 1980.

37. L. da F. Costa, F. Rodrigues, and G. Travieso. Characterization of complex networks: A survey of measurements. *Adv. Phys.*, 56:167–242, 2007.

38. C.M. da Fonseca, M. Ghebleh, A. Kanso, and D. Stevanovic. Counterexamples to a conjecture on Wiener index of common neighborhood graphs. *MATCH Commun. Math. Comput. Chem.*, 72:333–338, 2014.

39. Z. Daròczy and A. Jarai. On the measurable solutions of functional equation arising in information theory. *Acta Math. Acad. Sci. Hung.*, 34:105–116, 1979.

40. K.C. Das and M. Dehmer. A conjecture regarding the extremal values of graph entropy based on degree powers. *Entropy*, 18(5):1832016.

41. K.C. Das and Y. Shi. Some properties on entropies of graphs. *MATCH Commun. Math. Comput. Chem.*, 78:259–272, 2017.

42. M. Dehmer. Information processing in complex networks: Graph entropy and information functionals. *Appl. Math. Comput.*, 201:82–94, 2008.

43. M. Dehmer, N. Barbarini, K. Varmuza, and A. Graber. Novel topological descriptors for analyzing biological networks. *BMC Struct. Biol.*, 10:18, 2010.

44. M. Dehmer and F. Emmert-Streib, editors. *Quantitative Graph Theory—Mathematical Foundations and Applications*. CRC Press, New York, US, 2015.

45. M. Dehmer, F. Emmert-Streib, Z. Chen, X. Li, and Y. Shi, editors. *Mathematical Foundations and Applications of Graph Entropy*. Wiley, Weinheim, Germany, 2016.

46. M. Dehmer, F. Emmert-Streib, and M. Grabner. A computational approach to construct a multivariate complete graph invariant. *Inform. Sci.*, 260:200–208, 2014.

47. M. Dehmer, F. Emmert-Streib, B. Hu, Y. Shi, M. Stefu, and S. Tripathi. Highly unique network descriptors based on the roots of the permanental polynomial. *Inform. Sci.*, 408:176–181, 2017.

48. M. Dehmer, F. Emmert-Streib, and Y. Shi. Interrelations of graph distance measures based on topological indices. *PLoS ONE*, 9:e94985, 2014.

49. M. Dehmer, F. Emmert-Streib, and Y. Shi. Quantitative graph theory: A new branch of graph theory and network science. *Inform. Sci.*, 418:575–580, 2017.

50. M. Dehmer, F. Emmert-Streib, Y. Shi, M. Stefu, and S. Tripathi. Discrimination power of polynomial-based descriptors for graphs by using functional matrices. *PLoS ONE*, 10:e0139265, 2015.

51. M. Dehmer and A. Graber. The discrimination power of molecular identification numbers revisited. *MATCH Commun. Math. Comput. Chem.*, 69:785–794, 2013.

52. M. Dehmer and A. Ilić. Location of zeros of Wiener and distance polynomials. *PLoS ONE*, 7(3):e28328, 2012.

53. M. Dehmer and V. Kraus. On extremal properties of graph entropies. *MATCH Commun. Math. Comput. Chem.*, 68:889–912, 2012.

54. M. Dehmer, X. Li, and Y. Shi. Connections between generalized graph entropies and graph energy. *Complexity*, 21(1):35–41, 2015.

55. M. Dehmer and A. Mehler. A new method of measuring similarity for a special class of directed graphs. *Tatra Mt Math. Publ.*, 36:39–59, 2007.

56. M. Dehmer, M. Moosbrugger, and Y. Shi. Encoding structural information uniquely with polynomial-based descriptors by employing the randić matrix. *Appl. Math. Comput.*, 268:164–168, 2015.

57. M. Dehmer and A. Mowshowitz. Generalized graph entropies. *Complexity*, 17:45–50, 2011.

58. M. Dehmer and A. Mowshowitz. A history of graph entropy measures. *Inform. Sci.*, 1:57–78, 2011.

59. M. Dehmer, A. Mowshowitz, and F. Emmert-Streib. Connections between classical and parametric network entropies. *PLoS ONE*, 6:e15733, 2011.

60. M. Dehmer, A. Mowshowitz, and Y. Shi. Structural differentiation of graphs using Hosoya-based indices. *PLoS ONE*, 9(7):e1024592014.

61. M. Dehmer, L. Müller, and A. Graber. New polynomial-based molecular descriptors with low degeneracy. *PLoS ONE*, 5(7):e113932010.

62. M. Dehmer, Y. Shi, and A. Mowshowitz. Discrimination power of graph measures based on complex zeros of the partial Hosoya polynomial. *Appl. Math. Comput.*, 250:352–355, 2015.

63. M. Dehmer, L. Sivakumar, and K. Varmuza. Uniquely discriminating molecular structures using novel eigenvalue–based descriptors. *MATCH Commun. Math. Comput. Chem.*, 67:147–172, 2012.

64. M. Dehmer, K. Varmuza, and D. Bonchev, editors. *Statistical Modelling of Molecular Descriptors in QSAR/QSPR. Quantitative and Network Biology.* Wiley-Blackwell, New York, US, 2012.

65. M. Dehmer, K. Varmuza, S. Borgert, and F. Emmert-Streib. On entropy-based molecular descriptors: Statistical analysis of real and synthetic chemical structures. *J. Chem. Inf. Model.*, 49:1655–1663, 2009.

66. E. Deutsch and S. Klavžar. Computing the Hosoya polynomial of graphs from primary subgraphss. *MATCH Commun. Math. Comput. Chem.*, 70:627–644, 2013.

67. A. Dobrynin, R. Entringer, and I. Gutman. Wiener index of trees: Theory and applications. *Acta Appl. Math.*, 66:211–249, 2001.

68. A. Dobrynin and A.A. Kochetova. Degree distance of a graph: A degree analogue of the Wiener index. *J. Chem. Inf. Comput. Sci.*, 34:1082–1086, 1994.

69. S. N. Dorogovtsev and J. F. F. Mendes: *Evolution of Networks. From Biological Networks to the Internet and WWW.* Oxford University Press, Oxford, UK, 2003.

70. S. Dragomir and C. Goh. Some bounds on entropy measures in information theory. *Appl. Math. Lett.*, 10:23–28, 1997.

71. W. Du, X. Li, and Y. Shi. Algorithms and extremal problem on Wiener polarity index. *MATCH Commum. Math. Comput. Chem.*, 62(1):235–244, 2009.

72. Z. Duan, C. Liu, and G. Chen. Network synchronizability analysis: The theory of subgraphs and complementary graphs. *Phys. D*, 237(7):1006–1012, 2008.

73. N. Eagle, M. Macy, and R. Claxton. Network diversity and economic development. *Science*, 328:1029–1031, 2010.

74. J.A. Ellis-Monaghan and C. Merino. Graph polynomials and their applications i: The Tutte polynomial. In M. Dehmer, editor, *Structural Analysis of Complex Networks*, volume 4, pages 219–255. Springer, Berlin, 2011.

75. F. Emmert-Streib. The chronic fatigue syndrome: A comparative pathway analysis. *J. Comput. Biol.*, 14(7):961–972, 2007.

76. F. Emmert-Streib and M. Dehmer. Networks for systems biology: Conceptual connection of data and function. *IET Syst. Biol.*, 5(3):185–207, 2011.

77. F. Emmert-Streib, M. Dehmer, and Y. Shi. Fifty years of graph matching, network alignment and comparison. *Inform. Sci.*, 346–347:180–197, 2016.

78. R.C. Entringer, D.E. Jackson, and A.D. Snyder. Distance in graphs. *Czech. Math. J.*, 26:283–296, 1976.

79. P. Erdös. On the number of complete subgraphs contained in certain graphs. *Publ. Math. Inst. Hung. Acad. Sci.*, 7:459–464, 1962.

80. E. Estrada, L. Torres, L. Rodrǵuez, and I. Gutman. An atom-bond connectivity index: modelling the enthalpy of formation of alkanes. *Indian J. Chem.*, 37A:849–855, 1998.

81. J. Fang. From a single network to "networks of networks" development process: Some discussions on the exploration of multilayer supernetwork models and challenges. *Complex Syst. Complex. Sci.*, 13(1):40–47, 2016.

82. E. J. Farrell. An introduction to matching polynomials. *J. Comb. Theory B*, 27:75–86, 1979.

83. L. Feng, W. Liu, G. Yu, and S. Li. The hyper-Wiener index of graphs with given bipartition. *Utilitas Math.*, 95:23–32, 2014.

84. L. Feng, and G. Yu. The hyper-Wiener index of cacti. *Utilitas Math.*, 93:57–64, 2014.

85. S. Fortunato. Community detection in graphs. *Phys. Rep.*, 486(3–5):75–174, 2010.

86. L.C. Freeman. A set of measures of centrality based on betweenness. *Sociometry*, 40:35–41, 1977.

87. J. Gao, D. Li, and S. Havlin. From a single network to a network of networks. *Natl Sci. Rev.*, 1(3):346–356, 2014.

88. X. Gao, B. Xiao, D. Tao, and X. Li. A survey of graph edit distance. *Pattern Anal. Appl.*, 13(1):113–129, 2010.

89. C. D. Godsil and I. Gutman. On the theory of the matching polynomial. *J. Graph Theory*, 5:137–144, 1981.

90. V. Goldshtein, G.A. Koganov, and G.I. Surdutovich. Vulnerability and hierarchy of complex networks. *Cond. Mat.*, page 0409298, 2004.

91. I. Gutman. Acyclic systems with extremal Hückel π-electron energy. *Chem. Phys. Lett.*, 45:79–87, 1977.

92. I. Gutman. The energy of a graph. *Ber. Math. -Stat. Sekt. Forschungsz. Graz*, 103:1–22, 1978.

93. I. Gutman. A formula for the Wiener number of trees and its extension to graphs containing cycles. *Graph Theory Notes NY*, 27:9–15, 1994.

94. I. Gutman. Selected properties of the Schultz molecular topological index. *J. Chem. Inf. Comput. Sci.*, 34:1087–1089, 1994.

95. I. Gutman, B. Furtula, E. Zogić, and E. Glogić. Resolvent energy of graphs. *MATCH Commun. Math. Comput. Chem.*, 75:279–290, 2016.

96. I. Gutman and X. Li. *Energies of Graphs C Theory and Applications*. University of Kragujevac and Faculty of Science Kragujevac, Kragujevac, 2016.

97. I. Gutman, X. Li, and J. Zhang. Graph energy. In M. Dehmer and F. Emmert-Streib, editors, *Analysis of Complex Networks. From Biology to Linguistics*, pages 145–174. Wiley–VCH, Weinheim, 2009.

98. I. Gutman and N. Trinajstić. Graph theory and molecular orbitals. Total π-electron energy of alternant hydrocarbons. *Chem. Phys. Lett.*, 18:535–538, 1973.

99. I. Gutman and S. Wagner. The matching energy of a graph. *Discrete Appl. Math.*, 160:2177–2187, 2012.

100. I. Gutman and B. Zhou. Laplacian energy of a graph. *Linear Algebra Appl.*, 414:29–37, 2006.

101. F. Harary. *Graph Theory*. Addison Wesley Publishing Company, Boston, MA, 1969.

102. S. Harenberg, G. Bello, L. Gjeltema, S. Ranshous, J. Harlalka, R. Seay, K. Padmanabhan, and N. Samatova. Community detectionin large-scale networks: A surveyand empirical evaluation. *WIREs Comput. Stat.*, 6:426–439, 2014.

103. S.M. Hsieh and C.C. Hsu. Graph-based representation for similarity retrieval of symbolic images. *Data Knowl. Eng.*, 65(3):401–418, 2008.

104. D. Hu, X. Li, X. Liu, and S. Zhang. *Extremality of Graph Entropy Based on Degrees of Uniform Hypergraphs with Few Edges*. preprint.

105. B. Huo, S. Ji, X. Li, and Y. Shi. Solution to a conjecture on the maximal energy of bipartite bicyclic graphs. *Lin. Algebra Appl.*, 435:804–810, 2011.

106. B. Huo, X. Li, and Y. Shi. Complete solution to a conjecture on the maximal energy of unicyclic graphs. *Eur. J. Comb.*, 32:662–673, 2011.

107. A. Ilić. On the extremal values of general degree-based graph entropies. *Inform. Sci.*, 370C371:424C–427, 2016.

108. G. Indulal, I. Gutman, and A. Vijayakumar. On distance energy of graphs. *MATCH Commun. Math. Comput. Chem.*, 60:461–472, 2008.

109. O. Ivanciuc, T.S. Balaban, and A.T. Balaban. Reciprocal distance matrix, related local vertex invariants and topological indices. *J. Math. Chem.*, 12:309–318, 1993.

110. G. Jaklič, P.W. Fowler, and T. Pisanski. *hl*–index of a graph. *Ars Math. Contemp.*, 5:99–105, 2012.

111. Y. Jiang, B. Li, and J. Chen. Analysis of the velocity distribution in partially-filled circular pipe employing the principle of maximum entropy. *PLoS ONE*, 11(3), 2016.

112. M.R. Jooyandeh, D. Kiani, and M. Mirzakhah. Incidence energy of a graph. *MATCH Commun. Math. Comput. Chem.*, 62:561–572, 2009.

113. H.A. Jung. On a class of posets and the corresponding comparability graphs. *J. Combin. Theory Ser. B*, 24:125–133, 1978.

114. B.H. Junker and F. Schreiber. *Analysis of Biological Networks*. Wiley-Interscience, Berlin, 2008.

115. F. Kaden. Graph metrics and distance-graphs. In M. Fiedler, editor, *Graphs and Other Combinatorial Topics*, pages 145–158. Teubner Texte zur Math, Prague, Czech, 1983.

116. D.Y. Kenett, M. Perc, and S. Boccaletti. Networks of networks—An introduction. *Chaos Solit. Fractals*, 80:1–6, 2015.

117. L.B. Kier and L.H. Hall. *Molecular Connectivity in Chemistry and Drug Research*. Academic Press, New York, NY, 1976.

118. L.B. Kier and L.H. Hall. *Molecular Connectivity in Structure–Activity Analysis*. Wiley, New York, NY, 1986.

119. L.B. Kier and L.H. Hall. The meaning of molecular connectivity: A bimolecular accessibility model. *Croat. Chem. Acta*, 75:371–382, 2002.

120. M. Knor, B. Lužar, R. Škrekovski, and I. Gutman. On Wiener index of common neighborhood graphs. *MATCH Commun. Math. Comput. Chem.*, 72:321–332, 2014.

121. M. Knor, R. Škrekovski, and A. Tepeh. Mathematical aspects of Wiener index. *Ars Math. Contemp.*, 11:327–352, 2016.

122. E.V. Konstantinova. On some applications of information indices in chemical graph theory. In R. Ahlswede, L. Bäumer, N. Cai, H. Aydinian, V. Blinovsky, C. Deppe, and H. Mashurian, editors, *General Theory of Information Transfer and Combinatorics*, pages 831–852. Springer, Berlin, Germany, 2006.

123. J.H. Koolen and V. Moulton. Maximal energy graphs. *Adv. Appl. Math.*, 26:47–52, 2001.

124. J. Körner. Coding of an information source having ambiguous alphabet and the entropy of graphs. *Transactions of the 6-th Prague Conference on Information Theory*, pages 411–425, 1973.

125. J. Körner and V. Moulton. Maximal energy bipartite graphs. *Graphs Combin.*, 19:131–135, 2003.

126. J. Körner, V. Moulton, and I. Gutman. Improving the mcclelland inequality for total π-electron energy. *Chem. Phys. Lett.*, 320:213–216, 2000.

127. D. Koschützki, K.A. Lehmann, L. Peeters, S. Richter, D. Tenfelde-Podehl, and O. Zlotowski. Centrality indices. *Lect. Notes Comput. Sci.*, 3418:16–61, 2005.

128. V. Kraus, M. Dehmer, and F. Emmert-Streib. Probabilistic inequalities for evaluating structural network measures. *Inform. Sci.*, 288:220–245, 2014.

129. V. Latora and M. Marchiori. Vulnerability and protection of infrastructure networks. *Phys. Rev. E*, 71:015103(R), 2005.

130. X. Li and I. Gutman. *Mathematical Aspects of Randić-type Molecular Structure Descriptors*. University of Kragujevac and Faculty of Science Kragujevac, Kragujevac, Serbia, 2006.

131. X. Li, Y. Li, Y. Shi, and I. Gutman. Note on the HOMO-LUMO index of graphs. *MATCH Commun. Math. Comput. Chem.*, 70:85–96, 2013.

132. X. Li, Z. Qin, M. Wei, I. Gutman, and M. Dehmer. Novel inequalities for generalized graph entropies—Graph energies and topological indices. *Appl. Math. Comput.*, 259:470–479, 2015.

133. X. Li and Y. Shi. A survey on the Randić index. *MATCH Commun. Math. Comput. Chem.*, 59(1):127–156, 2008.

134. X. Li, Y. Shi, and I. Gutman. *Graph Energy*. Springer, New York, NY, 2012.

135. X. Li, Y. Shi, and L. Wang. An updated survey on the Randić index. In B. Furtula I. Gutman, editor, *Recent Results in the Theory of Randic Index*, pages 9–47. University of Kragujevac and Faculty of Science Kragujevac, Kragujevac, Serbia, 2008.

136. X. Li, P. Zhang, Z. Di, and Y. Fan. Community structure in complex networks. *Complex Syst. Complex. Sci.*, 5(3):19–42, 2008.

137. Y. Li, Y. Shi, and X. Gu. Scattering number, integrity, tenacity and the spectrum of regular graphs. *Future Gener. Comp. Syst.* 83:450–453, 2018.

138. H. Lin. Extremal Wiener index of trees with given number of vertices of even degree. *MATCH Commum. Math. Comput. Chem.*, 72:311–320, 2014.

139. H. Lin. On the Wiener index of trees with given number of branching vertices. *MATCH Commum. Math. Comput. Chem.*, 72:301–310, 2014.

140. J. Liu, B. Wang, and Q. Guo. Improved collaborative filtering algorithm via information transformation. *Int. J. Mod. Phys. C*, 20(2):285–293, 2009.

141. R. Liu, C. Jia, T. Zhou, D. Sun, and B. Wang. Personal recommendation via modified collaborative filtering. *Phys. A*, 388:462–468, 2009.

142. S. Liu and H. Zhang. On the characterizing properties of the permanental polynomials of graphs. *Linear Algebra Appl.*, 438:157–172, 2013.

143. G. Lu, B. Li, and L. Wang. Some new properties for degree-based graph entropies. *Entropy*, 17(12):2015.

144. G. Lu, B. Li, and L. Wang. New upper bound and lower bound for degree-based network entropy. *Symmetry*, 8(2):2016.

145. L. Lu, C. Jin, and T. Zhou. Similarity index based on local paths for link prediction of complex networks. *Phys. Rev. E*, 80(4):0461222009.

146. L. Lu and T. Zhou. Link prediction in complex networks: A survey. *Phys. A*, 390 (3):1150–1170, 2011.

147. L. Lu and T. Zhou. *Link Prediction (in Chinese)*. Higher Education Press, Beijing, 2013.

148. W. Lu, X. Li, and Z. Rong. Global stabilization of complex networks with digraph topologies via a local pinning algorithm. *Automatica*, 46(1):116–121, 2010.

149. J. Ma, Y. Shi, Z. Wang, and J. Yue. On wiener polarity index of bicyclic networks. *Sci. Rep.*, 6:19066, 2016.

150. J. Ma, Y. Shi, and J. Yue. The Wiener polarity index of graph products. *Ars Combin.*, 116:235–244, 2014.

151. F.D. Malliaros and M. Vazirgiannis. Clustering and community detection in directed networks: A survey. *Phys. Rep.*, 533(4):95–142, 2013.

152. B. J. McClelland. Properties of the latent roots of a matrix: The estimation of $\pi$-electron energies. *J. Chem. Phys.*, 54:640–643, 1971.

153. A. Mehler, P. Weiß, and A. Lücking. A network model of interpersonal alignment. *Entropy*, 12(6):1440–1483, 2010.

154. X. Meng, S. Sun, X. Li, L. Wang, C. Xia, and J. Sun. Interdependency enriches the spatial reciprocity in prisoner's dilemma game on weighted networks. *Phys. A*, 442:388–396, 2016.

155. B. Mohar. Median eigenvalues of bipartite planar graphs. *MATCH Commun. Math. Comput. Chem.*, 70:79–84, 2013.

156. B. Mohar. Median eigenvalues and the HOMO–LUMO index of graphs. *J. Combin. Theory B*, 112:78–92, 2015.

157. B. Mohar. Median eigenvalues of bipartite subcubic graphs. *Combin. Probab. Comput.*, 25:768–790, 2016.

158. B. Mohar and B. Tayfeh-Rezaie. Median eigenvalues of bipartite graphs. *J. Algebraic Combin.*, 40:899–909, 2015.

159. A. Mowshowitz. Entropy and the complexity of the graphs I: An index of the relative complexity of a graph. *Bull. Math. Biophys.*, 30:175–204, 1968.

160. M.E.J. Newman. Scientific collaboration networks. I. network construction and fundamental results. *Phys. Rev. E*, 64:016131, 2001.

161. M.E.J. Newman. The structure and function of complex networks. *SIAM Rev.*, 45 (2):167–256, 2003.

162. V. Nikiforov. The energy of graphs and matrices. *J. Math. Anal. Appl.*, 326:1472–1475, 2007.

163. D. Plavšić, S. Nikolić, N. Trinajstić, and Z. Mihalić. On the Harary index for the characterization of chemical graphs. *J. Math. Chem.*, 12:235–250, 1993.

164. M. Randić. On characterization of molecular branching. *J. Am. Chem. Soc.*, 97:6609–6615, 1975.

165. M. Randić. On molecular indentification numbers. *J. Chem. Inf. Comput. Sci.*, 24:164–175, 1986.

166. N. Rashevsky. Life, information theory, and topology. *Bull. Math. Biophys.*, 17:229–235, 1955.

167. P. Rényi. On measures of information and entropy. In *Proceedings of the 4th Berkeley Symposium on Mathematics, Statistics and Probability*, volume 1, pages 547–561. University of California Press, Berkeley, CA, 1961.

168. H.P. Schultz. Topological organic chemistry. 1. Graph theory and topological indices of alkanes. *J. Chem. Inf. Comput. Sci.*, 29:239–257, 1989.

169. C.E. Shannon and W. Weaver. *The Mathematical Theory of Communication.* University of Illinois Press, Urbana, IL, 1949.

170. D. Shi, Q. Chen, and L. Liu. Markov chain-based numerical method for degree distributions of growing networks. *Phys. Rev. E*, 71:036140, 2005.

171. Y. Shi, M. Dehmer, X. Li, and I. Gutman. *Graph Polynomials.* CRC Press, New York, NY, 2016.

172. R. Skrekovski and I. Gutman. Vertex version of the Wiener theorem. *MATCH Commun. Math. Comput. Chem.*, 72:295–300, 2014.

173. M.I. Skvortsova, I.I. Baskin, I.V. Stankevich, V.A. Palyulin, and N.S. Zefirov. Molecular similarity. 1. Analytical description of the set of graph similarity measures. *J. Chem. Inf. Comput. Sci.*, 38:785–790, 1998.

174. F. Sobik. Graphmetriken und klassifikation strukturierter objekte, ZKI-informationen. *Akad. Wiss. DDR*, 2:63–122, 1982.

175. A. Soltani, A. Iranmanesh, and Z. Abdul Majid. The multiplicative version of the edge Wiener index. *MATCH Commun. Math. Comput. Chem.*, 71:407–416, 2014.

176. S. Thurner. Statistical mechanics of complex networks, In M. Dehmer, and F. Emmert-Streib, *Analysis of Complex Networks: From Biology to Linguistics*, pages 23–45, Wiley-VCH, Berlin, Germany, 2009.

177. W.T. Tutte. A contribution to the theory of chromatic polynomials. *Can. J. Math.*, 6:80–91, 1954.

178. K. Varmuza and H. Scsibrany. Substructure isomorphism matrix. *J. Chem. Inf. Comput. Sci.*, 40:308–313, 2000.

179. G. Vivaldo, E. Masi, C. Pandolfi, S. Mancuso, and G. Caldarelli. Networks of plants: How to measure similarity in vegetable species. *Sci. Rep.*, 6:27077, 2016.

180. D. Vukičević and B. Furtula. Topological index based on the ratios of geometrical and arithmetical means of end-vertex degrees of edges. *J. Math. Chem.*, 46:1369–1376, 2009.

181. S. Wagner. Energy bounds for graphs with fixed cyclomatic number. *MATCH Commun. Math. Comput. Chem.*, 68:661–674, 2012.

182. C. Wang and A. Qu. Entropy, similarity measure and distance measure of vague soft sets and their relations. *Inform. Sci.*, 244:92–106, 2013.

183. X. Wang and G. Chen. Synchronization in scale-free dynamical networks: robustness and fragility. *IEEE Trans. Circuits Syst. I*, 49(1):54–62, 2002.

184. D. J. Watts and S. H. Strogatz. Collective dynamics of 'small-world' networks. *Nature*, 393:440–442, 1998.

185. H. Wiener. Structural determination of paraffn boiling points. *J. Am. Chem. Soc.*, 69:17–20, 1947.

186. W. Xia, J. Ren, F. Qi, Z. Song, M. Zhu, H. Yang, H. Jin, B. Wang, and T. Zhou. Empirical study on clique-degree distribution of networks. *Phys. Rev. E*, 76:037102, 2007.

187. K. Xu, M. Liu, K.C. Das, I. Gutman, and B. Furtula. A survey on graphs extremal with respect to distance-based topological indices. *MATCH Commun. Math. Comput. Chem.*, 71:461–508, 2014.

188. W. Yu, J. Cao, and J. Lu. Global synchronization of linearly hybrid coupled networks with time-varying delay. *SIAM J. Appl. Dyn. Syst.*, 7(1):2008.

189. B. Zelinka. On a certain distance between isomorphism classes of graphs. *Čas. Pest. Math.*, 100:371–373, 1975.

190. M. Zhao, G. Chen, T. Zhou, and B. Wang. Enhancing the network synchronizability. *Front. Phys. China*, 2(4):460–468, 2007.

191. M. Zhao, T. Zhou, B. Wang, G. Yan, and H. Yang. Effects of average distance and heterogeneity on network synchronizability. In *Proceedings of the 17th World Congress The International Federation of Automatic Control*, Seoul, Korea, 6–11, 2008.

192. L. Zhong. The harmonic index for graphs. *Appl. Math. Lett.*, 25:561–566, 2012.

193. B. Zhou, I. Gutman, J. A. de la Peña, J. Rada, and L. Mendoza. On the spectral moments and energy of graphs. *MATCH Commun. Math. Comput. Chem.*, 57:183–191, 2007.

194. B. Zhou and N. Trinajstić. On a novel connectivity index. *J. Math. Chem.*, 46:1252–1270, 2009.

195. T. Zhou, G. Yan, and B. Wang. Maximal planar networks with large clustering coeffcient and power-law degree distribution. *Phys. Rev. E*, 71(4):046141, 2005.

196. T. Zhou, M. Zhao, G. Chen, G. Yan, and B. Wang. Phase synchronization on scale-free networks with community structure. *Phys. Lett. A*, 368:431–434, 2007.

197. W. Zhou, N. Koptyug, S. Ye, Y. Jia, and X. Lu. An extended n-player network game and simulation of four investment strategies on a complex innovation network. *PLoS ONE*, 11:e0145407, 2016.

198. J. Petersen. Sur la théorème de Tait. *L'Intermédiare des Math.* 5:225–227, 1898.

# Chapter 5

## Overview of Social Media Content and Network Analysis

**Mohammed Ali Al-garadi, Henry Friday Nweke, and Ghulam Mujtaba**

**CONTENTS**

## 5.1 Introduction

Social media (SM) websites are supported by Web 2.0 technology which provides new feature that allows the users to create their own profiles and pages which makes the users more active. Whereas Web 1.0 limited the users to be only passive reader of the content, Web 2.0 expanded the capabilities which allow the users to interact with each other creating social network communication. SM have four particular potential capacities including collaboration, participation, empowerment, and time [1]. The interaction through the social networks (Facebook, Google+, LinkedIn, and Twitter) creates huge and useful amount of data. The social network data available online make it easy to study social phenomena with higher precision. This opportunity creates new interest to introduce social networks as link research between computer science and criminology, sociology, economy, and biological science which opens new and modern field of research in different research fields.

The SM websites contain social data posted by users which are used to comprehensively understand the human behaviors and society characteristics. This chapter gives an overview of online SM websites characteristics, content analysis method for SM, and network methods for SM.

## 5.2    Characteristics of Online SM Websites

The interaction through SM websites (e.g., Facebook, Google+, LinkedIn, and Twitter) produces vast and useful amounts of data. SM websites can be used as a platform for studying social phenomena with increased precision [2]. This opportunity creates new interest in introducing SM websites as a research link between computer science and criminology, sociology, economy, and biological science, which opens new and modern areas of research in various fields [3,4].

SM communication is a revolutionary trend that maximizes Web 2.0. American scientists Boyd and Ellison [5] describe Web 2.0 as "Web-based services that allow individuals to (1) construct a public or semi-public profile within a bounded system, (2) articulate a list of other users with whom they share a connection, and (3) view and traverse their list of connections and those made by others within the system."

*Presentation of oneself*: in this characteristic, a user in an SM through which he/she can introduce himself/herself to others creates a profile or account. The profile contains personal information, such as name, age, country, work, address, and interests. SM accounts are used to present the user through text, photos, music, and videos; the user can communicate with others by adding friends and contacts.

*Web address*: Users are given a unique web address that becomes their web identity. In this page or account, the users are provided space to upload, post, and share all of their data.

Social network has introduced new environment which people create their own ideas and knowledge then post and share it across their online friends' network and communities. People are exposed to various ideas, thoughts, cultures, and opinions, and hence, social network becomes a very important factor in building the society characteristic. A study from Brigham Young University was involved on 491 families, it studied the relationship between the teenagers and theirs parents in social network and it showed that the teenagers who communicate with their parents in social networks like Twitter and Facebook have a better connection with them in real life (offline). Parents can understand more about their teens' interests and friends [6]; furthermore, they can show affection positive feedback.

*User-generated data*: Users use SM websites to upload, post, and share text, images, and videos; this practice has produced a massive volume of user-generated data. The majority of SM users access SM sites every day [7]. They do not merely access the data available in an SM site; each individual spends a significant amount of time communicating and sharing their thoughts and emotions online. This phenomenon has resulted in the staggering growth of data available in SM websites.

*Interaction*: SM websites have created highly popular interactive environments for their users. Users can chat with one another, share ideas, comment on each other, post text on the pages of others, tag others in photos and videos, play games, and create events and invitations to these events.

*Creation of community*: SM websites allow the users to create and manage their own groups and pages; different communities with various interests are created in social networks. A social network such as Facebook allows members of the same group or page to share and post text, images, and videos, thereby creating virtual communities with real-life information.

*Popularity*: The popularity of SM websites is attributed to their simplicity. Basic Internet skills are sufficient to create and manage social network accounts. Prior to social networks, having webpages was limited because of technical and financial challenges. By contrast, the majority of SM websites are free of charge, open for anyone, and require simple steps of registration.

The connection characteristics of SM websites describe their structural features; they deal with how the users are linked with one another in the SM context. These connection characteristics can be categorized into large-scale network, dynamicity, and explicit and implicit connections.

*Large-scale network*: SM websites contain billions of users who are connected with one another through relational links; these links vary in each SM. For example, on Facebook, the link between connected users can be friendship link (bidirectional relationship) or following (one-directional relationship). On Twitter, these links are following and follower relationships (one-directional relationship). The vast numbers of users and links create large-scale networks and extensive human interactions. This phenomenon offers a unique opportunity for studying and understanding social interaction and communication among massive populations.

*Dynamicity*: SM websites are dynamic and continuously evolving as an increasing number of people are attracted to them. On a daily basis, numerous nodes (new users) join the network and huge numbers of links (relationships) are created between the nodes.

*Privacy and ethical issues*: SM websites introduced a new legal and ethical issue, the issue on the ownership of the data available on social network, and how to protect the copyright in user-generated data. Moreover sometimes, the messages contain sensitive information. Therefore, with privacy-preserving is an issue. Visibility of user's profile is necessary and it is an important factor in SM to introduce the users to new people. On the other hand, changing the privacy into public may cause attacks such as reputation slander and spamming

[8], and the Twitter users may avoid this by changing their privacy setting to private instead of public.

## 5.3    SM Analysis Methods

The SM analysis is divided mainly into content analysis and network. In the following sections, we briefly discuss the content and network of analysis of SM websites.

### 5.3.1    Content analysis of SM websites

SM websites have millions of users who post and update content daily, thereby generating a large amount of data called user-generated data. Mining SM involves the use of data mining technologies to find useful knowledge from social interaction data in social networks. Data mining of social networks can expand researchers' ability to understand new methods because of the wide use of SM for different purposes [9]. SM data, such as those from Facebook and Twitter, are collected through the API, which enables researchers and developers to read from and write data on Facebook and Twitter. The large amount of user-generated data in social networks can be analyzed with help of data mining approaches for many real-life applications.

#### 5.3.1.1    Data mining

Data mining is defined as the processes or steps of obtaining knowledge from huge amounts of data [10]. Data mining aims to recognize and define the patterns and correlations between data. Data mining is a relatively new field that has successfully created many methods and algorithms for analyzing big data for real-world application. Data mining processes involve either statistical or machine learning algorithms in the analysis stage to get the pattern from the big data of SM.

Data mining also equals knowledge discovery from data (KDD). Data mining and knowledge discovery process methodologies can be divided into three types [11]: knowledge discovery database and its related approaches; the Cross-Industry Standard Process for Data Mining (CRISP-DM) [12] as shown in Figure 5.1 and its related approaches; and other approaches.

KDD-related approaches involve the original KDD process by Fayyad et al. [13] and the approaches that originated from it directly, such as the approach developed by Anand et al. [14], SEMMA or sample, explore, modify, model, assess [15], and the method developed by Cabena et al. [16]. The nine steps are showed in the tables. there are as follows : Developing and (1) understanding of the application domain; (2) Creating a target data set; (3) Data cleaning and preprocessing; (4) Data reduction and projection; (5) Selecting the data mining function; (6) Selecting data mining algorithm; (7) Data mining; (8) Interpretation; (9) Discovered knowledge CRISP-DM [12] and its related approaches, such

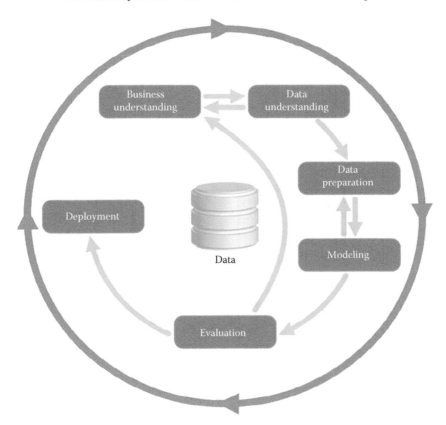

**FIGURE 5.1:** Process diagram showing the relationship between the different phases of data mining. (Adapted from P. Chapman et al., "CRISP-DM 1.0," *CRISP-DM Consortium*, 2000. [12])

as the process model of Cios and Kurgan [17] and the method of Marbán et al. [18,19], have six processes: field understanding, data understanding, data pre-processing, modeling, evaluation, and deployment. Other approaches include the KDD Roadmap [20] and Six Sigma, which was developed by Motorola in mid-1996 (Table 5.1) [21].

### 5.3.1.2   Machine learning algorithms

In recent years, machine learning algorithms have been extensively applied to understand the SM content [22–26]. Machine learning research has become an important task in many application areas and has magnificently created several methods and algorithms for analyzing the big data to solve real-world problems.

The most used machine learning algorithms to analyze the unstructured data from SM websites are supervised machine learning methods. In supervised learning algorithms, variables can be split into independent variables (test data)

**TABLE 5.1:**   List of different methodologies in data mining

Steps and phases →

| Methodology | No. of steps | | | | | | | | | |
|---|---|---|---|---|---|---|---|---|---|---|
| Fayyad et al. [13] | 9 | Developing and understanding of the application domain | Creating a target data set | Data cleaning and preprocessing | Data reduction and projection | Selecting the data mining function | Selecting data mining algorithm | Data mining | Interpretation | Discovered knowledge |
| CRISP-DM | 6 | Business understanding | Data understanding | Data preparation | Modeling | Evaluation | Deployment | | | |
| Anand et al. [14] | 8 | Human resource identification | Problem specification | Data prospecting | Domain knowledge elicitation | Methodology identification | Data preprocessing | Pattern discovery | Knowledge postprocessing | |
| Cabena et al. [16] | 5 | Business objectives determination | Data preparation | Data mining | Domain knowledge elicitation | Assimilation of knowledge | | | | |
| Cios and Kurgan [17] | 6 | Understanding the problem domain | Understanding the data | Preparation of the data | Data mining | Evaluation of the discovered knowledge | Using the discovered knowledge | | | |
| SEMMA | 5 | Sample | Explore | Explore modify | Model | Assess | | | | |
| Marbán et al. [18,19] | 6 | Selection | Transformation | Data mining | Preprocessing | Interpretation | Evaluation | | | |
| The five A's | 5 | Assess | Access | Analyze | Act | Automate | | | | |
| Six sigma (6-s) | 5 | Define | Measure | Analyze | Improve | Control | | | | |
| KDD Roadmap | 8 | Problem specification | Resourcing | Data cleansing | Reprocessing data | Data mining | Evaluation | | Interpretation | Exploitation |

and dependent variables (labeled training). An analysis is performed to find a relationship between the independent and the dependent variables. The model is built by using the labeled training of the data, which is then applied to test data to evaluate the accuracy of the model's predictions. For high effectiveness of supervised learning algorithm, the labeled training should be recognized for a large data set. Supervised learning is preferred when labeled training data are available. The typical supervised learning algorithms are classification algorithms, naïve Bayes, adaptive Bayes network, tree algorithms, support vector machines, K-nearest neighbors, and regression trees. Supervised machine leering algorithms have been extensively used to analysis SM websites data, for example, for detection crime [27], detecting cyberbullying [23], tracking the spread of infectious disease by analysis of SM data [23], financial prediction and marketing [28,29], election prediction [30], disaster prediction [31], and large human behavior studies [32,33].

### 5.3.2   SM network analysis

SM websites can be described as a structure that enables the exchange and dissemination of information, as well as social interaction between individuals [34]. SM websites have introduced an innovative way for scientists to study the structure of human relationships. A number of measurement and characterization studies on SM websites have attempted to provide a first step toward understanding the characteristics of SM websites; few such studies are based on complete datasets from SM operators [35,36]. In analyzing social network graphs, only part of the network structure is used because of the privacy regulations of SM websites, and directly acquiring the data from SM providers is difficult. Moreover, crawling millions of SM users is challenging [37]. In this section, we first discuss the crawling technique of SM networks. Secondly, we highlight the common complex network metrics that can be used for SM website network analysis.

#### 5.3.2.1   Crawling SM websites

SM websites can be represented as graphs, in which nodes are represent users, and edges represent connections. The first step in representing the network in a graph is to crawl its network structure. Crawling techniques are divided into (a) graph traversal techniques and (b) random walk.

#### 5.3.2.2   Graph traversals

Breadth-first search (BFS) is a graph traversal algorithm that is used to crawl SM websites. It shows high-order optimality and ease of implementation, particularly in undirected graphs. The working principle of BFS is as follows: the process begins with a seed node, and the neighbor nodes are explored. Then, all neighbors for each neighbor are explored. This process continues until the whole graph is explored and formed. BFS has been used to crawl Facebook [38], with the user friend list as the seed node. Then, the friend list of each user in the seed node is visited and so on. A parallel crawler framework for SM websites was created using BFS [37]. BFS has been adopted in many social network crawling efforts [35,39–41]. It collects a full graph (all nodes and links) of a particular region. However, BFS has shown bias toward high-degree nodes [41].

### 5.3.2.3   Depth-first search

Depth-First Search DFS is a graph traversal algorithm that traverses nodes systematically until its goal is found [42]. DFS is more in-depth than BFS, and it uses a last-in, first-out stack for storing unexpanded nodes. DFS is successful if it finds its goal. Otherwise, the loop will continue until success or failure is met. DFS requires less memory than BFS because it stores the nodes from root to current nodes.

### 5.3.2.4   Random walk

In random walks [43], the starting point is selected randomly. Then, its neighbor is selected randomly, moving to the selected neighbor, after which its neighbor is selected at random. This process will continue until the graph is represented. Random walk has been used in Twitter [44], Friendster [45], and Facebook [46]. Random walk leads to bias toward high-degree nodes, even though this bias can be corrected in certain degree distributions. However, random walk fails for large-scale graphs [47]. Reweighted random walk can be used to correct the bias in random graph [47], and this was achieved by using the Hansen–Hurwitz estimator [48].

### 5.3.2.5   Comparison between crawling algorithms for SM websites

In Table 5.2, comparison between crawling algorithms for SM websites is presented.

### 5.3.2.6   SM crawling challenges

In previous studies that analyzed SM websites, only part of the network structure is used because of the privacy regulations of SM websites and the difficulty of obtaining data directly from SM providers. Moreover, crawling millions of SM users is challenging [37]. Many SM websites use a number of dynamic pages, and users can easily modify their profiles and pages with high flexibility, thereby complicating the design and increasing the challenge for the crawler to handle such a complex network efficiently [41]. SM websites allow users to set their privacy. Some users prefer to keep their profile private and visible only to their friends. Thus, their profiles are black holes during extraction process. Most previous studies do not indicate how this condition impacts their results [41].

**TABLE 5.2:** Comparison between crawling algorithms for SM websites

| | BFS | DFS | Random walk |
|---|---|---|---|
| Advantages | • Easy to implement<br>• Optimal solution for unweighted graph<br>• Able to find a solution if it exists<br>• Able to find the solution with fewer steps when more than one solution exists<br>• It collects a full graph (all nodes and links) of a particular region. | • Lower memory requirement than BFS<br>• Able to find a solution without visiting all nodes, thereby requiring less time and space | • Simple to use<br>• Space saving<br>• Efficient<br>• Able to manage flows in complicated boundaries |
| Drawbacks | • Requires a large memory space<br>• Time consuming if the solution is farther away from the root<br>• Leads to bias toward high-degree nodes | • Cannot guarantee that a solution will be found<br>• Cannot guarantee that a solution will be found with minimal steps if more than one solution exists | • It does not conserve the nodes position on the space.<br>• Its solution is noisy because of statistical error.<br>• It leads to bias toward high-degree nodes even though this bias can be corrected in certain degree distributions; however, it fails for large-scale graphs [47]. Reweighted random walk can be used to correct the bias in random graph [47], which was achieved by using the Hansen–Hurwitz estimator [48]. |

### 5.3.2.7    SM websites metrics (measurement)

SM websites are an example of systems constructed by a huge number of highly interconnected nodes (users). An SM can be denoted as graphs, the links connect the nodes where the nodes are the users, and the links are the relationships between the users in the SM. Therefore, detailed investigation and understanding of SM websites are essential to comprehend the relationship among people and assist in responding to several queries about humanity and sociality at large-scale analysis. To understand such network, many complex network techniques have been proposed [40]. These measurements such as degree centrality, betweenness centrality closeness centrality k-core, and PageRank are commonly used for identifying vital nodes in networks which those nodes can be used for viral marketing [23,49–52].

## 5.4    Conclusion

SM users grow rapidly and they are encouraged post, communicate, and get socialized with friends with the network the data available through social network can give us insights into social communities and societies. The SM websites contain social data posted by users which explain the human behaviors and society characteristics. This chapter gives an overview of online SM websites characteristics, content analysis method for SM, and network methods for SM.

## References

1. M. J. Magro, "A review of social media use in e-government," *Administrative Sciences*, vol. 2, no. 2, pp. 148–161, 2012.

2. J. Ratkiewicz, M. Conover, M. R. Meiss, B. Gonçalves, A. Flammini, and F. Menczer, "Detecting and tracking political abuse in social media," *ICWSM*, vol. 11, pp. 297–304, 2011.

3. H. Lauw, J. C. Shafer, R. Agrawal, and A. Ntoulas, "Homophily in the digital world: A LiveJournal case study," *Internet Computing, IEEE*, vol. 14, no. 2, pp. 15–23, 2010.

4. M. Dehmer, F. Emmert-Streib, A. Mehler, J. Kilian, "Measuring the structural similarity of web-based documents: A novel approach," *International Journal of Computational Intelligence*, vol. 3, no. 1, pp. 1–7, 2006.

5. D. M. Boyd and N. B. Ellison, "Social network sites: Definition, history, and scholarship," *Journal of Computer-Mediated Communication*, vol. 13, no. 1, pp. 210–230, 2007.

6. S. M. Coyne, L. M. Padilla-Walker, R. D. Day, J. Harper, and L. Stockdale, "A friend request from dear old dad: Associations between parent–child social networking and

adolescent outcomes," *Cyberpsychology, Behavior, and Social Networking*, vol. 17, no. 1, pp. 8–13, 2014.

7. C. M. Cheung, P.-Y. Chiu, and M. K. Lee, "Online social networks: Why do students use Facebook?," *Computers in Human Behavior*, vol. 27, no. 4, pp. 1337–1343, 2011.

8. G. Hogben, "Security issues and recommendations for online social networks," *ENISA Position Paper*, no. 1, 2007.

9. P. Gundecha and H. Liu, "Mining social media: A brief introduction," *Tutorials in Operations Research*, vol. 1, no. 4, pp. 22–23, 2012.

10. J. Han, J. Pei, and M. Kamber, *Data Mining: Concepts and Techniques*. Elsevier, 2011.

11. G. Mariscal, Ó. Marbán, and C. Fernández, "A survey of data mining and knowledge discovery process models and methodologies," *Knowledge Engineering Review*, vol. 25, no. 2, pp. 137, 2010.

12. P. Chapman et al., "CRISP-DM 1.0," *CRISP-DM Consortium*, 2000.

13. U. M. Fayyad, G. Piatetsky-Shapiro, P. Smyth, and R. Uthurusamy, *Advances in Knowledge Discovery and Data Mining*. American Association for Artificial Intelligence, Menlo Park, CA, 1996.

14. A. G. Buchner, M. D. Mulvenna, S. S. Anand, and J. G. Hughes, "An internet-enabled knowledge discovery process," in *Proceedings of the 9th International Database Conference*, Hong Kong, 1999, pp. 13–27.

15. SAS Institute, "Semma Data Mining Methodology," 2005.

16. P. Cabena, P. Hadjinian, R. Stadler, J. Verhees, and A. Zanasi, *Discovering Data Mining: From Concept to Implementation*. Prentice Hall, Upper Saddle River, NJ, 1998.

17. K. J. Cios and L. A. Kurgan, "Trends in data mining and knowledge discovery," in *Advanced Techniques in Knowledge Discovery and Data Mining*. Springer, London, 2005, pp. 1–26.

18. Ó. Marbán, G. Mariscal, E. Menasalvas, and J. Segovia, "An engineering approach to data mining projects," in *Intelligent Data Engineering and Automated Learning*. Springer, Berlin, Heidelberg, 2007, pp. 578–588.

19. O. Marbán, J. Segovia, E. Menasalvas, and C. Fernández-Baizán, "Toward data mining engineering: A software engineering approach," *Information Systems*, vol. 34, no. 1, pp. 87–107, 2009.

20. J. C. W. Debuse, B. de la Iglesia, C. M. Howard, and V. J. Rayward-Smith, "Building the KDD Roadmap," in *Industrial Knowledge Management*. Springer, London, 2001, pp. 179–196

21. T. Pyzdek and P. A. Keller, "The six sigma handbook," vol. 4, McGraw-Hill Education, New York, NY, 2014.

22. H. Schoen, D. Gayo-Avello, P. Takis Metaxas, E. Mustafaraj, M. Strohmaier, and P. Gloor, "The power of prediction with social media," *Internet Research*, vol. 23, no. 5, pp. 528–543, 2013.

23. M. A. Al-garadi, K. D. Varathan, and S. D. Ravana, "Cybercrime detection in online communications: The experimental case of cyberbullying detection in the Twitter network," *Computers in Human Behavior*, vol. 63, pp. 433–443, 2016.

24. S. Zerdoumi, A. Q. Md Sabri, A. Kamsin, I. A. T. Hashem, A. Gani, S. Hakak, M. A. Al-garadi, V. Chang, "Image pattern recognition in big data: Taxonomy and open challenges: Survey," *Multimedia Tools and Applications*, pp. 1–31, 2017.

25. G. Bello-Orgaz, J. J. Jung, and D. Camacho, "Social big data: Recent achievements and new challenges," *Information Fusion*, vol. 28, pp. 45–59, 2016.

26. E. Olshannikova, T. Olsson, J. Huhtamäki, and H. Kärkkäinen, "Conceptualizing Big Social Data," *Journal of Big Data*, vol. 4, no. 1, p. 3, 2017.

27. X. Wang, M. S. Gerber, and D. E. Brown, "Automatic crime prediction using events extracted from Twitter posts," *SBP*, vol. 12, pp. 231–238, 2012.

28. J. Bollen, H. Mao, and X. Zeng, "Twitter mood predicts the stock market," *Journal of Computational Science*, vol. 2, no. 1, pp. 1–8, 2011.

29. A. Mittal and A. Goel, "Stock prediction using twitter sentiment analysis," *Standford University, CS229* (2011 http://cs229.stanford.edu/proj2011/GoelMittal-StockMarketPredictionUsingTwitterSentimentAnalysis.pdf), vol. 15, 2012.

30. A. Tumasjan, T. O. Sprenger, P. G. Sandner, and I. M. Welpe, "Election forecasts with Twitter: How 140 characters reflect the political landscape," *Social Science Computer Review*, vol. 29, no. 4, pp. 402–418, 2011.

31. H. Li, N. Guevara, N. Herndon, D. Caragea, K. Neppalli, C. Caragea, A. C. Squicciarini, and A. H. Tapia, "Twitter mining for disaster response: A domain adaptation approach," in *ISCRAM*, 2015.

32. V. Subrahmanian and S. Kumar, "Predicting human behavior: The next frontiers," *Science*, vol. 355, no. 6324, pp. 489–489, 2017.

33. D. Ruths and J. Pfeffer, "Social media for large studies of behavior," *Science*, vol. 346, no. 6213, pp. 1063–1064, 2014.

34. M. Dehmer, F. Emmert-Streib, and A. Zulauf, A graph mining technique for automatic classification of web genre data. na, 2007.

35. Y.-Y. Ahn, S. Han, H. Kwak, S. Moon, and H. Jeong, "Analysis of topological characteristics of huge online social networking services," in *Proceedings of the 16th International Conference on World Wide Web*, 2007, pp. 835–844: ACM.

36. J. Leskovec, L. Backstrom, R. Kumar, and A. Tomkins, "Microscopic evolution of social networks," in *Proceedings of the 14th ACM SIGKDD International Conference on Knowledge Discovery and Data Mining*, 2008, pp. 462–470: ACM.

37. D. H. Chau, S. Pandit, S. Wang, and C. Faloutsos, "Parallel crawling for online social networks," in *Proceedings of the 16th International Conference on World Wide Web*, 2007, pp. 1283–1284: ACM.

38. S. A. Catanese, P. De Meo, E. Ferrara, G. Fiumara, and A. Provetti, "Crawling Facebook for social network analysis purposes," in *Proceedings of the International Conference on Web Intelligence, Mining and Semantics*, 2011, p. 52: ACM.

39. C. Wilson, B. Boe, A. Sala, K. P. Puttaswamy, and B. Y. Zhao, "User interactions in social networks and their implications," in *Proceedings of the 4th ACM European Conference on Computer Systems*, 2009, pp. 205–218: ACM.

40. A. Mislove, M. Marcon, K. P. Gummadi, P. Druschel, and B. Bhattacharjee, "Measurement and analysis of online social networks," in *Proceedings of the 7th ACM SIG-COMM Conference on Internet Measurement*, 2007, pp. 29–42: ACM.

41. S. Ye, J. Lang, and F. Wu, "Crawling online social graphs," in *Web Conference (APWEB), 2010 12th International Asia-Pacific*, 2010, pp. 236–242: IEEE.

42. M. Tsvetovat and A. Kouznetsov, *Social Network Analysis for Startups: Finding Connections on the Social Web*, O'Reilly Media, Inc., 2011.

43. L. Lovász, "Random walks on graphs: A survey," *Combinatorics, Paul Erdos is Eighty*, vol. 2, no. 1, pp. 1–46, 1993.

44. B. Krishnamurthy, P. Gill, and M. Arlitt, "A few chirps about twitter," in *Proceedings of the First Workshop on Online Social Networks*, 2008, pp. 19–24: ACM.

45. A. H. Rasti, M. Torkjazi, R. Rejaie, and D. Stutzbach, "Evaluating sampling techniques for large dynamic graphs," *Univ. Oregon, Tech. Rep. CIS-TR-08*, vol. 1, 2008.

46. M. Gjoka, M. Kurant, C. T. Butts, and A. Markopoulou, "Walking in Facebook: A case study of unbiased sampling of OSNS," in *INFOCOM, 2010 Proceedings IEEE*, 2010, pp. 1–9: IEEE.

47. M. Gjoka, M. Kurant, C. T. Butts, and A. Markopoulou, "A walk in Facebook: Uniform sampling of users in online social networks," arXiv preprint arXiv:0906.0060, 2009.

48. M. J. Salganik and D. D. Heckathorn, "Sampling and estimation in hidden populations using respondent-driven sampling," *Sociological Methodology*, vol. 34, no. 1, pp. 193–240, 2004.

49. M. A. Al-garadi, K. D. Varathan, and S. D. Ravana, "Identification of influential spreaders in online social networks using interaction weighted K-core decomposition method," *Physica A: Statistical Mechanics and its Applications*, vol. 468, pp. 278–288, 2017.

50. M. S. Khan, A. W. A. Wahab, T. Herawan, G. Mujtaba, S. Danjuma, and M. A. Al-Garadi, "Virtual community detection through the association between prime nodes in online social networks and its application to ranking algorithms," *IEEE Access*, vol. 4, pp. 9614–9624, 2016.

51. S. Pei, L. Muchnik, J. S., Andrade Jr, Z. Zheng, and H. A. Makse, "Searching for superspreaders of information in real-world social media," *Scientific Reports*, vol. 4, pp. 5547, 2014.

52. J. Weng, E.-P. Lim, J. Jiang, and Q. He, "Twitterrank: Finding topic-sensitive influential twitterers," in *Proceedings of the Third ACM International Conference on Web Search and Data Mining*, 2010, pp. 261–270: ACM.

# Chapter 6

## Analysis of Critical Infrastructure Network

David Rehak, Pavel Senovsky, and Martin Hromada

### CONTENTS

## 6.1 Introduction

Since time immemorial, society has profited from infrastructures providing the civilian population with access to specific services. The issue of the security of these infrastructures has gradually become part of a broader discussion around the security of society, especially in light of the lessons learned from World War II and, subsequently, the concerns arising from the ensuing Cold War.

The rapid technological development after World War II and, in particular, the arrival of information technology (IT), have had a major impact on the way society functions. Society has become increasingly dependent on these infrastructures. IT contributed to enhancing the efficiency of managing these infrastructures and played a pivotal role in interconnecting them. At the same time, however, these significant changes also gave rise to a number of new challenges in the form of new, until then unknown, vulnerabilities capable of threatening the society as such. Eventually, the term "critical infrastructure" (CI) was adopted to describe these infrastructures.

With advancing urbanization, the CI system has both increased in its significance and become more vulnerable. Its individual elements are routinely exposed to the adverse effects of various threats of anthropogenic or naturogenic origin. It is, therefore, crucial to know the structure not only of the CI network, but also of its individual elements, for which resilience to the impacts of the emergencies has to be developed and maintained.

The main contribution of this chapter is in description of state of the art in the field of the complex networks forming CI of the state, problems and approaches to solve them from safety and security point of view.

## 6.2    Critical Infrastructure Network

Although the term CI is now widely used, its origins are not entirely clear. The person credited for using the term for the first time in public [6] was Charles C. Lane during the U.S. Senate hearing on the vulnerability of telecommunications and energy sources to terrorism. Its first mention in legislation dates back to 1996, when it was used as part of Executive Order EO 13010 on CI Protection [19]. The EO also established the Presidents Commission on Critical Infrastructure Protection (PCCIP)[1].

During its relatively brief existence, the Commission released a comprehensive report titled Critical Foundations [35], summarizing the current condition of CI and recommending measures to achieve a higher level of CI protection and improved resilience. The report provided the basis for the further development in CI protection, for the preparation of strategic documentation and for a reconsideration of relevant laws in the USA and, over time, also in numerous other countries.

The EO 13010 [19] defines CI as systems "*so vital that their incapacity or destruction would have a debilitating impact on the defence or economic security of the United States*". Similar definition has subsequently been adopted by other countries.

---

[1]The inception of CI has been described in great detail in book critical path [6].

Figure 6.1, a crude oil and petroleum products chart, has been provided as an example.

Even though the general definition of CI is widely accepted, the infrastructures included by different countries in the CI system vary considerably. A recent comparison of different approaches is available in the ENISA study [11]; see also Table 6.1 or ETH [7].

Despite the significant differences in the sectoral approach, the same elements can be traced between individual countries. For instance, the majority of countries agree that the CI is not only composed of physical networks (e.g., the rail system), but also includes civil administration. Accordingly, the CI should be viewed as an interaction between three principal stakeholders:

- CI owners/operators,

- public administration, and

- the public (society as a whole).

At the same time, each group has different demands and expectations with respect to the CI. CI owners are normally private businesses running the CI for profit. The public view the CI as a crucial precondition for the functioning of society. Public administration is concerned with market regulation (in the event of market failures, public administration as such can, on account of its law enforcement and governance functions, also be regarded as a CI component).

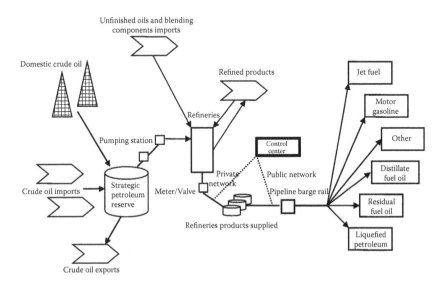

**FIGURE 6.1:** Crude oil and petroleum products processing and distribution system chart. (Courtesy of Spellman and Bieber [45].)

**TABLE 6.1:**   CI sectors of choice in different countries

| Sectors | Energy | ICT | Water | Food | Financial | Transport | Chemical |
|---|---|---|---|---|---|---|---|
| AU | X | X | X | X | X | X | |
| BE | X | X | | | X | X | |
| CZ | X | X | X | X | X | X | |
| DK | X | X | | X | | X | |
| EE | X | X | X | X | X | X | |
| FI | X | X | X | X | X | X | |
| FR | X | X | X | X | X | X | |
| DE | X | X | X | X | X | X | |
| EL | X | | | | | X | |
| HU | X | X | X | X | X | X | |
| IT | X | | | | | X | |
| MT | X | X | | | X | X | |
| NL | X | X | X | X | X | X | X |
| PL | X | X | X | X | X | X | X |
| SK | X | X | X | | | X | |
| ES | X | X | X | X | X | X | X |
| UK | X | X | X | X | X | X | |
| CH | X | X | X | X | X | X | |

*Source:* Courtesy of ENISA [11].

### 6.2.1   Critical infrastructure network system

The CI networks represent complex systems with strong intra- and inter-sectoral linkages. In order to obtain a complete picture of CI behavior and the consequences of potential failure, its individual elements or sectors must not be viewed separately, but rather in an integral manner. However, this approach is exceedingly complex and, at present, there is no method in place that would cover the issue in all its complexity.

Considering the CI network as such, it can be defined [25] as follows:

$$G(t) = \{N(t), L(t), f(t) : J(t)\} \tag{6.1}$$

where $t =$ time; $N =$ node of network (vertex of graph); $L =$ link of network (edge of graph); $f =$ network topology specification; $J =$ algorithm describing the behavior of nodes and links in a network.

The network structure interpretation (Equation 6.1) varies from network to network. For example, in a road network, the nodes consist of intersections or cities and the links are represented by the roads connecting them. With respect to air transport, the network nodes are the airports and the links between them are the air routes used.

The majority of CI networks are quite extensive in that they are composed of an increasing number of nodes. The analysis of CI and other extensive networks is the subject of *Network science*. Network science is a discipline devoted to the study of *complex networks*. This includes CI networks, and also social, biological, and various other networks.

Although the foundations of network theory were laid as early as in the eighteenth century, the study of complex networks was only made possible by research into random network models [12], small-world networks, in particular [50], and scale-free networks [1].[2] This discipline is relatively young and, as such, fraught with challenges. The 2005 study for the National Research Council [30] identified numerous challenges, some of which are especially relevant with respect to CI networks:

- Existing gaps in understanding the relationship between network architecture and its function.

- The modeling and analysis of particularly extensive networks is still a major challenge due to the heavy demands placed on the quality and amount of information on network.

- The lack of techniques facilitating network design or modification to ensure the network exhibits specific characteristics.

- Inadequate understanding of network robustness and security requirements.

There is a multitude of various perspectives and approaches that address failures and their modeling. For example, Perrow [37] has formulated the *normal accidents* theory. According to this theory, major failures/accidents have usually the same causes as minor ones. Perrow approaches the issue through the lens of a sociologist—his argument is based on human error and the inability of organizations to respond in time to prevent further escalation of accidents. "Normal" accidents are so-called by Perrow because such accidents are inevitable in complex systems.

Even though the theory was first put forward in 1984, it continues to be very well applicable to CI systems. Perrow identifies three conditions that make a system likely to be susceptible to Normal Accidents: the system is complex, the system is tightly coupled, and the system has catastrophic potential. All three conditions also aptly describe CI systems. From a system functionality perspective, these attributes are typical for systems in a position of Self-Organized Criticality (SOC).

For such systems, Bak et al. [4] argued that system destabilization, including the exact location of its occurrence and the magnitude of its consequences, is not possible to predict even for extremely simple situations.

The self-organization principle is very important as it allows the acceptance of certain assumptions about the topology of the networks. A pioneer in this area

---

[2]For a more detailed history of network science development, see [25].

was Barabasi [5], who analyzed a network of computers connected to the Internet and observed that this network was not random, as had until then been assumed, but constituted a scale-free network.

The difference between random and scale-free networks is clearly illustrated in Figure 6.2.

The existing research (i.e., [26]) is consistent with the view that most (if not all) CI networks are scale-free in terms of their topology. The link degree distribution between nodes in a scale-free network follows a power law, at least asymptotically.

$$P(k) \sim k^{-\gamma} \tag{6.2}$$

where $P(k)$ = the likelihood that a node is in the neighborhood of other network nodes; $\gamma$ = the distribution coefficient, usually $\geq 1$. The network $\gamma$ coefficient must be inferred from an analysis of real networks.

These types of networks characteristically tend to create hubs combining a great number of nodes. There usually are connections between individual hubs within a network. From the perspective of CI networks, some basic properties can be derived from the above observations:

- The distribution of nodes within a CI network is not random.

- The random removal of a node from the network is unlikely to significantly impact the network as a whole.

- The nonrandom organization of the network will increase its vulnerability if high-degree nodes are deliberately removed from the network.

Deliberate attempts to damage an infrastructure cannot be ruled out or dismissed due to the expected significant impacts this would have on the population and the economy. A CI thus appears to be an attractive target for terrorists or hostile nations.

In connection with this, terms such as *cyber weapons* or *cyber warfare* are becoming increasingly common, although there is currently no uniform opinion

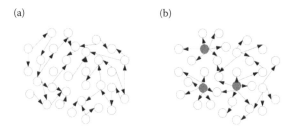

**FIGURE 6.2:** Random versus scale-free networks. (Courtesy of Castillo [8].)

among experts on this particular issue. For example, Rid [40] views "cyber warfare" as a contradiction in terms and argues that warfare operations generally involve a political aspect in the form of an unambiguous identification of the parties to the conflict and their political objectives in the conflict. However, countries that sponsor cyber attacks are currently capitalizing on the anonymity that the cyber environment provides. Rid, therefore, suggests that this issue falls into the sphere of intelligence operations.

An analysis of some of the most recent attacks can help form an idea of the tools (weapons) useful for such attacks and of their destructive potential (see, Stuxnet [51], botnet Mirai [16,23], etc.).

In terms of impacts on the CI network, the existence of hubs may lead to the disintegration of the network split as a consequence of a node being removed or incapacitated. Network breakdown may lead either to the failure of the entire network or the unavailability of certain services it provides. In the case of power grids, a power outage would be a typical consequence, whereas the damage to the Internet network would likely be limited to the unavailability of parts of the network, with its basic function preserved.

In view of the above, it is not possible to predict which particular element of the CI system is likely to fail. However, based on the characteristics of the CI network (see, e.g., [27]) and its individual elements, it is possible, to a certain extent, to estimate the general trajectory of such failures; for example, the frequency and size of outages in the infrastructures of concern, as well as the ease with which such failure would propagate throughout the network (a condition known as the epidemic threshold).

This type of information provides a sound basis for risk management and resilience development in these infrastructures over the long term.

### 6.2.2   Interconnected critical infrastructure networks

CI systems usually do not function independently, but tend to form complex conglomerates of interconnected infrastructures. A problem in the network of one infrastructure can thus spread to other infrastructures via the existing linkages between these networks.

The characteristics of these linkages or interdependencies can vary [42]:

- Physical—the condition of the infrastructure is dependent on the performance of another infrastructure.

- Cyber—reliance on information transmission (e.g., guidelines) via an information infrastructure.

- Geographical—for example, the collocation of the network elements of different CI infrastructures.

- Logical.

The network models in individual CI sectors are directly applicable to the management of these networks. In order to capture inter-sectoral linkages, it is necessary to use different type of models than those used to analyse the behavior of networks of individual infrastructures of interest. A good example of such a model is the Input–Output inoperability model [44].

These types of models make it possible, to a certain extent, to quantify the impacts of infrastructure failures on the networks of other (interconnected) infrastructures, and thereby facilitate understanding of the character of infrastructure linkages and their categorization for risk management purposes.

In addition to CI protection, *critical infrastructure resilience* is also of particular interest. The issue of CI resilience becomes vital in view of the simple fact that CI networks (e.g., the power grid) are generally too extensive to ensure complete protection [35] of all their elements. The focus accordingly shifts from security objectives to developing the ability of these networks to maintain the highest possible level of services despite adverse external or internal influences (gracefully degrade) and to recover their function in the shortest possible time (bounce back). The term *resilience* has become standard in describing these abilities.

## 6.3    Resilience of Critical Infrastructure

The term resilience was first defined by Holling [18] in connection with the resistance and stability of ecological systems (later socioecological systems). Over time, the term began to appear in other scientific fields, including sociology, psychology, and economics. The youngest field, with respect to system resilience research, is engineering.

In the area of ecological systems, Holling [18] distinguishes two types of system behavior. The first type, stability, is the ability of a system to return to an equilibrium state after a temporary disturbance, and the more rapidly it returns, the more stable it is. The second type of system behavior, known as resilience, is a measure of the ability of a system to absorb impacts while maintaining the system's function. At the same time, it also holds true that resilient systems can oscillate considerably around the equilibrium state; thus, resilient systems are not necessarily stable systems.

In order to grasp the complexity of economic, ecological, and social systems, it is necessary to factor in the hierarchy and the adaptive cycles occurring within these systems. These complex processes are known as panarchy [15]. Panarchy can be defined as a structure in which natural systems and human ones are interlinked in an infinite adaptive cycle of restructuralization and recovery.

Research into the resilience of socioecological systems has also sparked an interest in research focused on resilience in society. The following definition, provided by the United Nations [47], is regarded as the one that is most commonly used: "*The ability of a system, community or society exposed to hazards to resist, absorb, accommodate, adapt to, transform and recover from the effects of a hazard in a timely and efficient manner, including through the preservation and restoration of its essential basic structures and functions through risk management*. Yet other authors argue that the resilience of a society is proportionate to its ability to respond to a specific stress factor and can also be defined as the ability of groups or communities to cope with external stress situations and failures brought about by societal, political, or environmental changes [22].

Resilience gradually began to be defined in general terms for any system, including engineering ones, where the focus of attention is on the issue of CI. For example, resilience according to Walker et al. [49] represents the capacity of a system to absorb disturbance and reorganization, whereas it undergoes a change so that it continues to maintain essentially the same function, structure, identity, and feedback. With respect to enhancing robustness/resistance, emphasis is placed on system dynamics where the system has deviated far from its state of equilibrium.

Resilience was first described in connection with CI in a document entitled Critical Infrastructure Resilience: Final Report and Recommendations [31], where it is defined as the ability to absorb, adapt to, and/or rapidly recover from a potentially disruptive event. By contrast, the Critical Infrastructure Resilience Strategy [3] defines CI resilience as the *"ability to reduce the magnitude and/or duration of a disruptive event"*.

These definitions clearly show what constitutes resilience, or rather what characteristics enhance the resilience of a system. For example, Chandra [9], based on his study of socioecological systems, includes the following attributes in engineering systems resilience: redundancy, adaptability, flexibility, interoperability, and diversity.

## 6.3.1   Concept of critical infrastructure resilience

Based on the accepted definitions, resilience can be said to represent the level of internal preparedness of CI subsystems for adverse events, or the ability of these subsystems to perform and maintain their functions when negatively affected by internal and/or external factors. Enhancing resilience (e.g., [24,38]) minimizes the vulnerability of subsystems, which in turn curtails the occurrence, intensity, and propagation of failures and their impacts in a CI system and society. During emergencies, resilient systems actually show a smaller decrease in performance, and the time needed for them to return to their required level of operation is measurably shorter.

Understanding and a clear definition of resilience represent the cornerstone of resilience assessment and strengthening with respect to CI subsystems. In fact, CI system resilience must be understood as a cyclic process based on continual enhancement of prevention, absorption, recovery, and adaptation (see Figure 6.3).

The first phase of the CI resilience cycle is *prevention*. By adopting preventive measures, the owner/operator ensures that the system is less vulnerable to future emergencies. Once an emergency occurs, such a system will switch from prevention to absorption.

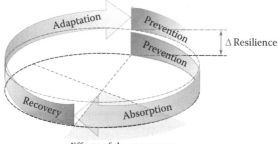

Effects of the emergency

**FIGURE 6.3:** Cycle of CI Resilience.

*Absorption*, the second phase of the resilience cycle, is initiated if a subsystem is disrupted due to an emergency, and is determined by the CI subsystem's robustness. Accordingly, robustness is defined as the ability of a CI element to absorb the effects of an emergency without the occurrence of any fluctuations in the services provided by the infrastructure concerned. In a CI system, two types of robustness are recognized: structural and security robustness. Structural robustness is determined by the progress of the decline of a function and the level of redundancy, while security robustness is based on the level of protective measures, detection, and ability to respond.

The *recovery* phase starts after the effects of an emergency have worn off. This phase is characterized by restorability, which is the capacity of a subsystem to restore its function to the required level of performance after the effects of an emergency no longer exist. The success of the restoration is determined by the available resources and the time required to complete the restoration process.

The final phase of the CI resilience cycle is *adaptation*, which is essentially the ability of an organization to adapt a subsystem to the effects resulting from an emergency. It represents the dynamic long-term ability of an organization to adapt to changes in circumstances. Adaptation is determined by the internal processes of an organization focused on the strengthening of resilience, that is, risk management and innovation/education processes. However, the strengthening of subsystem resilience already occurs in the phase of recovery of its performance.

### 6.3.2    Variables determining the resilience of critical infrastructure elements

The resilience of elements in a CI system lies in two basic areas, the first of which involves the technological and physical protection of individual elements. This type of resilience, known as technical resilience, is determined by the robustness and recoverability of system elements. The enhancement of technical resilience is invariably achieved exclusively in relation to a particular element or group of identical or very similar elements. A good example is the electricity sector, where robustness and recoverability will be secured in different ways and by different means depending on whether we are dealing with systems for the production of electricity or systems employed for its transmission and distribution. The second area is organization management. This type of resilience, known as organizational resilience, is determined by the level of an organization's internal processes whose core purpose is to create optimum conditions for the adaptation of CI elements to emergencies.

*Technical resilience* is determined by the robustness and recoverability of CI elements. For each element, these two components are determined or influenced by three basic factors: the technological structure of the element, the security measures of the element, and the scenario of the emergency for which resilience has been established. See Figure 6.4 for the components and their variables defined for the establishment of technical resilience.

*Robustness* is the ability of an element to absorb the impacts of an emergency. These impacts may be absorbed via the technologies used (i.e., structural robustness) and/or via security measures (i.e., security robustness). In general, robustness is determined by the level of crisis preparedness (i.e., safety/security plans, protective measures, progress of decrease of function, see Figure 6.5), the level or redundancy (i.e., the ability to substitute parts of the element or expand its capacity promptly), detection, and ability to respond. The level of robustness can only be assessed relative to a particular emergency or, rather, the intensity thereof. Where this level reaches 100%, the element concerned becomes resistant to the impacts of the given emergency.

*Recoverability* is the capacity of an element to recover its function to the original (required) level of performance after the effects of an emergency

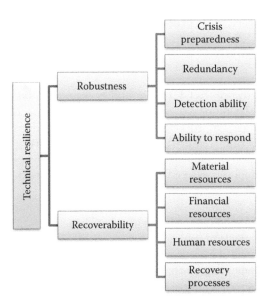

**FIGURE 6.4:** Technical resilience components and variables.

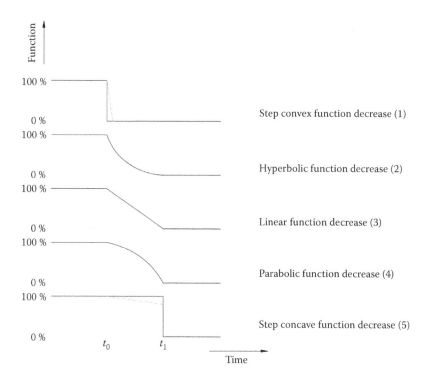

**FIGURE 6.5:** Progress of the decrease of a CI elements function. (Adapted from Lovecek et al. Determining the resilience of transport critical infrastructure element: Use case. *In Proceedings of 21st International Conference Transport Means 2017*, 21, 824–828, Kaunas University of Technology, Kaunas, Lithuania, 2017. [28])

have ended. With respect to CI, recoverability is understood as reparability, in which case only the damaged or destroyed components of an element are repaired or replaced. The success of recovery is determined by the available resources and the time required to complete the recovery process. If these resources are adequate, resilience can already be enhanced at this stage. The implementation of more modern technologies, ensuring greater element robustness, can be used as an example.

*Organizational resilience* is formed simultaneously for all CI elements operated by an organization. This type of resilience is formed, assessed, and enhanced by organization management as early as in the prevention phase, and factors in the level of internal processes necessary for the phase of *adaptation* of CI elements to emergencies that have already occurred. These processes are risk management, innovation processes, and education/development processes. See Figure 6.6 for the processes and their variables defined for the establishment of organizational resilience.

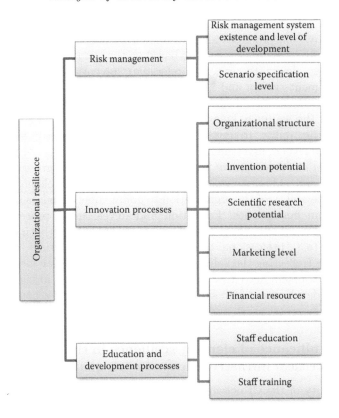

**FIGURE 6.6:** Organizational resilience processes and variables.

*Risk management* is a significant internal process of an organization that is essential to ensuring security/safety and enhancing resilience at the prevention stage. Risk management consists of coordinated activities to direct and control an organization with regard to risk (ISO 31000 [21]), and its level in relation to organizational resilience is determined by two basic criteria. The first criterion is the existence and level of a risk management system, that is, the level of adoption and utilization of the risk management strategy, risk analysis, risk control, risk monitoring, and risk management optimization. The second criterion is the degree of specification of emergency scenarios, which form the very basis for developing contingency plans.

Additional internal processes which materially contribute to enhancing the resilience of CI elements at the stage of prevention are *organization innovation processes*. From a strictly practical point of view, innovation is categorized into product, process, marketing, and organizational innovation (Oslo Manual [32]). With regard to resilience enhancement, process, and organizational innovations are especially important, as they focus on the reliability and external security of the technologies used. The innovation process itself consists of three basic phases, which are invention, science and research, and implementation. The levels of innovation processes are determined by five basic criteria: organizational structure (a dynamic structure based on the openness of communication and processes, flexibility, etc., is preferable), invention potential (e.g., an idea management system and interdisciplinary cooperation), science-research potential (i.e., the ability of science-research workers, the level of implemented projects and the way new knowledge is transferred into practice), marketing level (e.g., the ability to implement all proposed changes), and financial resources (the level of investments, financial reserves, etc.).

*Education and development* processes constitute the last group of processes which form and enhance organizational resilience of CI elements, thereby improving the ability of the organization to adapt these elements to the effects resulting from emergencies. Education and development processes can be divided into three basic categories [2] which are knowledge (both explicit and tacit), skills (e.g., professional-technical, managerial, analytical, conceptual), and attitudes (reflecting the values held by an individual). Key forms of education and development activities include long-term education, foreign study programs, skills development (soft skills), professional training (for both preventive and emergency procedures), and staff training.

### 6.3.3    Significance of element resilience in the context of critical infrastructure network quality

The quality of the CI network is constantly affected by numerous factors which can be classified as internal and external factors. The most significant internal factors include:

- Sector and sub-sector network topology (an optimum complex network is one with an adequate number of nodes).

- Node distribution (an optimum distribution is one where nodes are arranged based on the need for the services a CI network provides).

- Node resilience (an optimum state is high resilience it must be noted here that the primary concern in this regard should be the resilience of the network as a whole and not just of its individual elements, since the CI network is essentially a system of systems[3]).

- Subsystem linkages, that is, elements, subsectors and sectors (in general, lower-intensity linkages can be said to be less likely to transmit cascading impacts when dealing with an emergency).

---

[3]ISO/IEC/IEEE 15,288—A system of systems (SoS) brings together a set of systems for a task that none of the systems can accomplish on its own. Each constituent system has its own management, goals, and resources while coordinating with other systems within the SoS and adapting to meet SoS goals.

The most important negative external factors affecting the network quality of CI are threats. These threats can be political (e.g., threat of war or hostile regimes), economic (e.g., budgetary or inflation threats), social (e.g., health or criminal threats), technological (e.g., energy or cyber threats), legal (e.g., litigation threats), or environmental (e.g., natural disasters or climatic changes). Conversely, the most significant positive external factor is financial support stimulating investment in boosting the performance and resilience of CI elements.

It is clear from the above that resilience is one of the key factors influencing the quality and, primarily, the security of CI networks. Its significance is further emphasized by the following outcomes, achieved through the high resilience of individual elements:

- Delivery of the required performance of CI elements.

- Minimization of the propagation of cascading and synergistic effects of emergencies throughout the CI system; that is, reduction of the speed at which a failure can spread in the network.

- Minimization of the necessary investment in network recovery following an emergency.

- Mitigation of the impacts of emergencies on society.

- Enhancement of the resilience of the entire CI network.

## 6.4  Cascading and Synergistic Effects of Critical Infrastructure Network Elements

A significant vulnerability of CI network elements lies in the high potential of emergencies to spread via cascading and synergistic effects. In recent years, these terms have increasingly been used in a wider context; however, their meaning is not the same. The terminological definition of a cascading and synergistic effect is often associated with various areas and activities of society. In the context of major industrial accident prevention, the domino effect is defined as the potential for an increased likelihood of the occurrence of major accidents, or the aggravated consequences thereof, due to territorial linkages between structures housing hazardous materials (SEVESO III [13]). Therefore, the cascading effect can be represented by continuous system changes that occur in succession and are brought about by one or more initiative events. See Figure 6.7 for a graphical representation of this process.

A negative change is essentially an adverse impact on the system as a whole. However, the consequences of cascading effects can also be argued to have a positive influence on the system as a whole (such as network operating in island

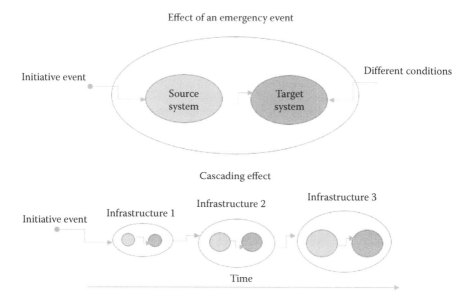

**FIGURE 6.7:** Cascading effect. (Adapted from J. Markuci. *Determining the synergic effect of national critical infrastructure [thesis].* VSB Technical University of Ostrava, Ostrava, Czech Republic, 2014. [29])

regime as result of cascading electrical power outages, which can be considered more robust and resilient way to operate network, provided that the network actually safely can degrade to island regime).

In order to better grasp the idea of a synergistic effect, it may be useful to look at the health sector, where the synergistic effect is related to the assessment of the cumulative effects of pharmaceuticals [10]. In the chemical industry, this phenomenon is associated with the use of catalytic agents to increase the intensity of a reaction [14]. In the area of management it can be most notably encountered in the context of profit maximization. An obvious parallel to the issue of CI can be found in connection with major industrial accident prevention, where the synergistic effect is associated with the establishment of emergency planning zones and the evaluation of synergistic impacts of hazardous materials or substances during emergency events [29].

From a terminological point of view, it can be said that due to the simultaneous action and coaction of multiple factors and agents within network linkages, synergistic effects exert qualitatively and quantitatively greater impacts than the regular sum of impacts of these factors. See Figure 6.8 for a comparison and graphical representation of the issue concerned.

The cascading effect can be said to create a succession of sorts, or rather a scenario of a system event development where, conversely, the synergistic effect predicts the level of impacts caused by these event development scenario [39].

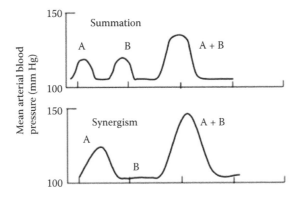

**FIGURE 6.8:** Comparison of the impact of drugs on blood pressure. (Adapted from W.C. Clarkson. *Basic Principles of Pharmacology*, 2016. [10])

### 6.4.1 Analysis of cascading and synergistic effects in a critical infrastructure network

An analysis of emergencies impacts points to the fact that, within the context of cascading and synergistic effects, multiple phenomena act simultaneously as the emergency develops [48].

The conceptual expression of cascading and synergistic effects is often linked to the issue of CI vulnerability and at present time is not dealt with in a comprehensive and systematic way. This failing is related to the phenomena of indirect impacts, collective harm, trading loss, and the destruction of cultural values and the environment. For a complex system, which CI certainly is, there is no coherent model that would effectively address this issue [41].

Consequently, the cascading effect is viewed as the successive failure of several elements which spreads to other parts of the system via network linkages. CI is obviously a complex system consisting of an extensive array of elements that interact and function as a whole. As a result, this can produce a synergistic effect [42].

Synergy, therefore, determines the degree of impacts from the degradation of network element functions in a CI, which increase over time as a result of the cascading effect. The level of the potential increase in impacts is determined by the criticality and intensity of the linkage to other CI elements.

In connection with the issue in question, the synergistic effect can be expected to occur as a consequence of a continual degradation of the function of two or more elements. This means that the interlinkages existing between system elements are the source and determine the occurrence of the cascading as well as the synergistic effect and the associated impacts.

The following section will, therefore, outline potential approaches to the assessment of network element interlinkages in terms of their potential relation to cascading and synergistic effects.

### 6.4.1.1    Interdependency matrix

The interdependency matrix philosophy is also applied to the assessment and analysis of the correlation between system elements, including in the context of the qualitative risk correlation analysis [34]. The application of similar methods is viewed in terms of assessment of interdependencies between individual network elements and the possibility of assessing the degree of influence of each element or subsector and its dependence on other CI subsectors. In connection with the research conducted by Pederson et al. [36], the interdependencies were discretely assessed using matrices of interdependencies between selected sectors [29,36], see Table 6.2 for a possible result.

The use of interdependency matrices can be regarded as a basis for the assessment of impacts of the synergistic effect in interdependent and interconnected sectors and network elements of CI. This method makes it possible to identify whether there is a linkage between selected elements and, at the same

**TABLE 6.2:**    Matrix of simple interdependencies

| Sector | Element | Energy and Utilities | | | | | Services | | |
|---|---|---|---|---|---|---|---|---|---|
| | | Electrical power | Water purification | Sewage treatment | Natural gas | Oil industry | Customs and imigration | Hospital and health care services | Food industry |
| Energy and Utilities | Electrical Power | | L | | | M | | | |
| | Water Purification | H | | | | M | | | |
| | Sewage Treatment | M | H | | | H | | | |
| | Natural Gas | L | | | | L | | | |
| | Oil Industry | H | L | | | | | | |
| Services | Customs & Immigration | H | L | L | L | L | | L | |
| | Hospital and Health Care Services | H | H | L | H | H | M | | H |
| | Food Industry | H | H | H | L | M | M | L | |
| | Key: H = High, M = Medium, L = Low | | | | | | | | |

*Source:* Adapted from Pederson et al. *Critical Infrastructure Interdependency Modeling: A Survey of U. S. and International Research.* INL, Idaho Falls, ID, 2006. [36]

time, determine the intensity of any such linkage in a subjective and qualitative manner. This is reflected in the comprehensive approach to the assessment of synergistic effect impacts resulting from the failure of a selected CI subsector in the Czech Republic [29].

### 6.4.1.2 Aggregate supply and demand model

This model falls into the category of tools evaluating the demand for CI services within a territorial unit and the ability to provide such services to the required extent. It is clear that as a networked system, CI creates a wider spectrum of demand for services, partly as a result of its uniqueness and inter-connectedness. The ability or, as the case may be, inability of a selected element to provide and distribute the required service or function points to its resilience and/or vulnerability.

Accordingly, this model should allow for the practical evaluation of the impact of the cascading effect brought about by the unavailability of required functions. This model formed the basis for the development and application of the theory of networks, nodes and edges in a directed graph. Within this context, the edges model the flow of commodities between nodes and may represent, for example, power transmission lines, water pipelines and roads, while the nodes represent selected CI elements [17]. Figure 6.9 shows factors representing processes that enter into network infrastructure nodes, that is, factors of time, the transfer of commodities, consumption, and production [17].

When applied to the simulation process, the model in question can, for example, help express the relationship between multiple utility systems and the water-supply network. In view of the above, the development in the flow of services and functions can be said not to be constant and, as such, can be simulated in the context of network performance from the outset of its disruption to its recovery [29].

### 6.4.1.3 Monte Carlo method

The Monte Carlo method can be used for simulating the anticipated values and standard deviations from the anticipated demand within a network containing multiple nodes, that is, CI elements that consume the given service or function.

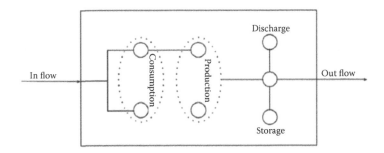

**FIGURE 6.9:** Matrix of simple interdependencies. (Adapted from Holden et al. *Saf. Sci.*, 53(Suppl. C):51–60, 2013. [17])

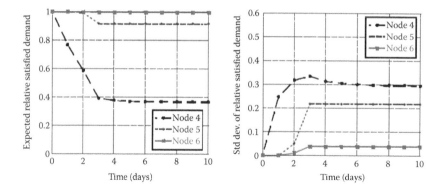

**FIGURE 6.10:** Relative satisfied electricity demand (left: expected value, right: standard deviation). (Adapted from Holden et al. *Saf. Sci.*, 53(Suppl. C):51–60, 2013. [17])

The results of the possible application of the selected method, in the context of accommodating the demand of individual consumers, are presented in Figure 6.10.

The graphs clearly show the way in which the selected method can assess the potential impacts caused by linkages between CI network elements. These impacts represent and are related to the time flow of the element malfunction.

### 6.4.1.4 Dynamic function modeling

The application of dynamic function modeling, aimed at determining the function of CI in the context of the effect of degradation and associated consequences, is principally based on the generation, distribution, and consumption of selected functions by other infrastructure systems. One of its advantages lies in the objective assessment of the influence exerted by politics and/or the legal, normative, and institutional framework with respect to the operation and direction of CI activities. The potential and applicability of this type of modeling were addressed in selected publications, where models of the vulnerability and ability to cooperate (interaction) of critical network infrastructures were developed based on the linkages between vulnerable nodes and threats with potential impacts on functionality. This approach was presented in publications [43,46], where all vulnerable nodes were defined as a major functional component of CI, fulfilling the demand for functions and services at local and regional levels. The threat node is characterized by an entity with a real potential for causing the disruption or degradation of a vulnerable node or a network element of CI. The threat node is subsequently expressed by the type of disruption or emergency in relation to the vulnerable node. In general, a model based on threat nodes can be said to pursue two basic approaches [29]:

1. The extent of the threats impact is independent of the specific condition of the affected node.

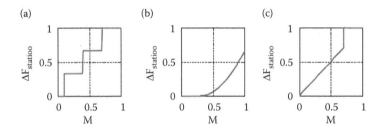

**FIGURE 6.11:** Example of functional integrity recovery. (Adapted from Trucco et al. *Reliab. Eng. Syst. Saf.*, 105(Suppl. C):51–63, 2012. [46])

2. If multiple threats have a common anticipated impact on a vulnerable node, the overall effect will be evaluated as the sum of impacts likely to be caused by each individual threat.

Therefore, threat distribution and propagation can be outlined using a functional integrity recovery function, as shown in Figure 6.11.

a. An example of the modulation function in the case of a car crash on a three-lane highway: depending on the seriousness of the event, the rescuers may decide to close one lane, two lanes, or the whole carriageway, so that the modulation function has a step form.

b. An example of the functional integrity modulation form in the case of a natural disaster, such as a flood: the CI does not suffer from low-level inundations; when the magnitude crosses a certain threshold, the higher the flood, the higher the impact on the CI.

c. An example in the case of a fire: for a small fire, the higher the magnitude, the higher the impact; for a larger fire, the CI is completely blocked.

The analysed publication refers to the source of disruption as the "father node", whereas the "child node" represents the node affected by the disrupted function. For graphical representation of dynamic modeling, see Figure 6.12.

The presented dynamic modeling was subsequently put into practice in Thailand, where the applicability of the selected model was amply demonstrated in terms of its ability to represent a wider set of interdependencies. The model covered 169 vulnerable nodes, 420 functional linkages, and 229 logical linkages. The key output was the formulation of the relationship between the characteristics of the node vulnerability ranging from its functional integrity to its transient damage via calculation of the degree of the malfunction the level of current demand, the changes in demand and the maximum level of provided services [46].

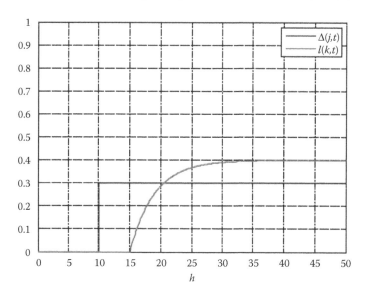

**FIGURE 6.12:** The relationship between the damaged "father node" $\Delta(j,t)$ and the malfunction of the "child node" $l(k,t)$. (Adapted from Senovsky et al. *Critical Infrastructure Protection*. SPBI, Ostrava, 2007. [43])

### 6.4.2   Determining the principle of modeling the impacts of critical network infrastructure failures

The internal description of the model factors in the relationships between system elements, their response to the input $u$ and their current condition $x$. If the system is linear, the changes in its condition can be described by an equation:

$$\frac{dx}{dt} = Ax + Bu \tag{6.3}$$

or, in the case of a discontinuous system, by a formula:

$$x(k+1) = Ax(k) + Bu(k) \tag{6.4}$$

Leontief's economic model, one of the applicable models, can also be employed in the context of critical network infrastructure, where $x$ represents the degradation (due to the cascading and synergistic effect) of CI network elements, $u$ is the input value causing the initial degradation of CI network elements, $A$ is the pairwise dependence matrix whose elements describe the pair interdependence between two elements and reflect the transfer of degradation from one element to the other. Let us express the paired relationship between elements $X_i$ and $X_j$ graphically, $X_j \rightarrow X_i$, with the transfer coefficient $a_{i,j}$.

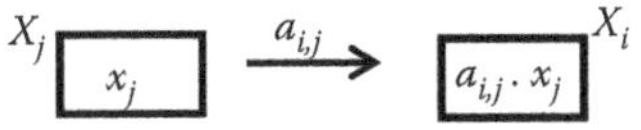

$$X_j \boxed{x_j} \xrightarrow{a_{i,j}} \boxed{a_{i,j} \cdot x_j}^{X_i}$$

**FIGURE 6.13:** First degree transfer. (Adapted from Hromada et al. *System and Method of Assessing Critical Infrastructure Resilience*. SPBI, Ostrava, 2013. [20])

If the degradation value of element $X_j$ is $x_j$, the value of its transfer to element $X_i$ will be $a_{i,j} \cdot x_j$. This constitutes a first degree transfer (Figure 6.13).

The value $a_{i,j} \cdot x_j$ spreads from element $X_j$ to other elements, whereby a second degree transfer is created. A subsequent transfer will create a third degree transfer and so on. The following figure (Figure 6.14) shows the pair interdependence of five CI elements, which defines the first degree transfer.

Second degree transfers include $X_1 \rightarrow X_2 \rightarrow X_3$, $X_1 \rightarrow X_2 \rightarrow X_4$, $X_2 \rightarrow X_3 \rightarrow X_5$, $X_2 \rightarrow X_4 \rightarrow X_5$.

Third degree transfers include $X_1 \rightarrow X_2 \rightarrow X_3 \rightarrow X_5$, $X_1 \rightarrow X_2 \rightarrow X_4 \rightarrow X_1$.

For example, if element $X_1$ were to be degraded, its degradation would increase as a result of the $X_1 \rightarrow X_2 \rightarrow X_4 \rightarrow X_1$ transfer [20].

The indicated transfers only represent the potential ways in which system degradation can spread. What matters is which element has been degraded, that is, which element is active (degraded) and from which element the degradation will spread. Figure 6.15 shows the initial degradation of the first element with the value of $X_1 = 0.15$ (active element).

From there, the degradation will spread via a first degree transfer to the second element with the value of $0.15 \cdot 0.2 = 0.03$. While the degradation does not transfer to other elements, the element designated as "active" with respect to the subsequent transfer will be the second element (Figure 6.16).

From the second element, the degradation with the value of $0.03 \cdot 0.3 = 0.009$ will spread via second degree transfer to the third element and thereafter to the fourth element with the value of $0.03 \cdot 0.5 = 0.015$. The "active" elements are $X_3$ and $X_4$ (Figure 6.17).

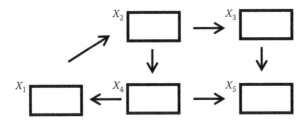

**FIGURE 6.14:** Third degree degradation. (Adapted from Hromada et al. *System and Method of Assessing Critical Infrastructure Resilience*. SPBI, Ostrava, 2013. [20])

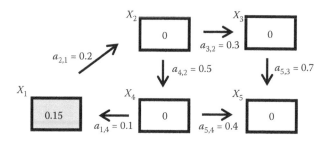

**FIGURE 6.15:** First degree degradation. (Adapted from Hromada et al. *System and Method of Assessing Critical Infrastructure Resilience.* SPBI, Ostrava, 2013. [20])

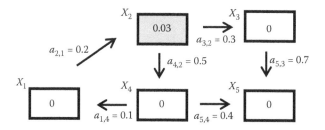

**FIGURE 6.16:** Second degree degradation. (Adapted from Hromada et al. *System and Method of Assessing Critical Infrastructure Resilience.* SPBI, Ostrava, 2013. [20])

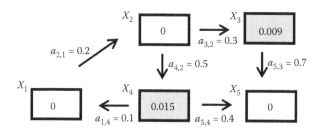

**FIGURE 6.17:** Degradation to the third element. (Adapted from Hromada et al. *System and Method of Assessing Critical Infrastructure Resilience.* SPBI, Ostrava, 2013. [20])

From the third and fourth elements, third degree degradation is transferred to $X_5$ and $X_1$ elements with the value of $0.009 \cdot 0.7 + 0.015 \cdot 0.4 = 0.0123$ to $X_5$ and $0.015 \cdot 0.1 = 0.0015$ to $X_1$, respectively. The active elements are $X_1$ and $X_5$; however, with no transfer occurring from $X_5$, the entire cycle is repeated with the active $X_1$ element only (Figure 6.18).

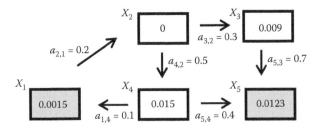

**FIGURE 6.18:** Third degree degradation. (Adapted from Hromada et al. *System and Method of Assessing Critical Infrastructure Resilience.* SPBI, Ostrava, 2013. [20])

In general, the entry of $u$ into the system will cause the initial degradation of its elements, which then spreads through their linkages to the remaining elements, producing a cascading and synergistic effect whereby the degradation propagates throughout the system. If $x_0 = u$ is the initial degradation, it will extend via pairwise dependence to subsequent degradation, increased by $\Delta 1 = Ax_0$, which will in turn produce an alteration of $\Delta 2 = A\Delta 1 = A(Ax_0) = A_2 x_0$, and so on. Following $n$-steps, an additional increase of $\Delta n = A(\Delta n - 1) = Anx_0$ will occur. As with Leontief's economic model, the question is whether the transfer of degradation will stabilize at the final value $x$, which will no longer vary, so that equilibrium will be reached. This will be the case if the resulting value of the degradation $x$ is broken down into the sum of the initial degradation $u$ and the degradation produced by the cascading and synergistic effect. This is represented by the equation where $x = Ax + u$. However, in this case, the sums in each column of the matrix $A$ may be greater than 1 and the equation may not have a solution. If an equilibrium occurs, the resulting degradation will be $x = (I - A) - 1u$ [33].

In the context of the network character of CI, system functionality and resilience were defined as the response of the output function $y$ to the input $u$, which is the final consumer perspective. This can be described in greater detail from an internal point of view, that is, via state dynamics analysis, which allows for final output corrections through system control, that is, management of state changes through additional inputs. Inputs should be viewed in two ways: as undesirable inputs degrading the system, and as inputs which mitigate or reduce the degradation (control inputs or measures). However, it may not always be easy to determine the impact a given input will have on system degradation.

## 6.5 Conclusion

CI networks are complex systems with strong intra- and inter-sectoral linkages. The majority of CI networks are quite extensive in that they are composed of an increasing number of nodes. Based on the existing research, the majority of CI networks (if not all) can be said to be scale-free in terms of their topology. These types of networks characteristically tend to create hubs combining a great number of nodes. There are usually connections between individual hubs within a network. From the perspective of CI networks, some basic characteristics can be derived: (1) node distribution within a CI network is not random; (2) the random removal of a node from the network is unlikely to significantly impact the network as a whole; and (3) nonrandom organization of the network will increase its vulnerability if key nodes are deliberately removed from the network. In terms of impacts on the CI network, the existence of hubs may eventuate in the disintegration of the network as a consequence of a node being removed or incapacitated. Network breakdown may lead either to the failure of the entire network or the unavailability of certain services it provides to the society.

It is, therefore, imperative that the resilience of a CI system be enhanced on a continuous basis, as it enables the system to absorb, adapt to, and/or rapidly recover from a potentially disruptive event. CI system resilience must be understood as a cyclic process based on continual enhancement of prevention, absorption, recovery, and adaptation. Current research clearly shows a close link with the quality and, above all, the security of CI networks. The enhancement of resilience of individual elements helps to (1) improve the required performance of CI elements; (2) minimize the propagation of cascading and synergistic effects of emergencies throughout the CI system; (3) minimize investment in network recovery following an emergency; (4) mitigate the impacts of emergencies on society; and (5) enhance resilience of the entire CI network.

Network characteristics and the interdependencies between individual CI subsystems alter the current understanding of the vulnerability of CIs with respect to the evaluation and improvement of their security. The complexity of the network infrastructure linkages indicated herein can thus be expected to continue to present challenges into the future.

## Acknowledgements

Writing of this chapter has been supported by the grant of the Ministry of the Interior of the Czech Republic VI20152019049—Dynamic Resilience Evaluation of Interrelated Critical Infrastructure Subsystems.

## References

1. R. Albert and A.-L. Barabasi. Statistical mechanics of complex networks. *Rev. Mod. Phys.*, 74(1):47–97, 2002.

2. M. Armstrong. *Armstrong's Handbook of Human Resource Management Practice.* Kogan Page, London, 13 edition, 2014.

3. Australian Government. *Critical Infrastructure Resilience Strategy.* Commonwealth of Australia, Cannbera, 2010.

4. P. Bak, C. Tang and K. Wiesenfeld. Self-organized criticality—An explanation of 1/F noise. *Phys. Rev. Lett.*, 59(4):381–384, 1987. WOS:A1987J291900001.

5. A.-L. Barabasi and J. Frangos. *Linked: The New Science of Networks Science of Networks.* Basic Books, New York, NY, 2014.

6. K. A. Brown. *Critical Path.* Spectrum Publishing Group, Inc., Fairfax, 2006.

7. E. M. Brunner and M. Suter. *International CIIP Handbook 2008/2009.* ETH, Curich, 2009.

8. C. Castillo. *Effective Web Crawling.* University of Chile, Santiago de Chile, 2004.

9. A. Chandra. *Synergy between Biology and Systems Resilience.* Missouri University of Science and Technology, Rolla, MO, 2010.

10. W. C. Clarkson. *Basic Principles of Pharmacology.* Tulane University, New Orleans, LA, 2016.

11. ENISA. *Methodologies for the Identification of Critical Information Infrastructure Assets and Services Guidelines for Charting Electronic Data Communication Networks.* ENISA, 2014.

12. P. Erdos and A. Renyi. On random graphs I. *Publ. Math. Debrecen,* 6:290–297, 1959.

13. European Parliament and Council. *Directive 2012/18/EU on the control of major-accident hazards involving dangerous substances, amending and subsequently repealing Council Directive 96/82/ EC (SEVESO III).*

14. J. Frimmel. Catalytic hydrolysis of chlorinated compounds on sulfide catalysts. *Chem. Listy,* 91(10):840–845, 1997.

15. L. H. Gunderson and C. S. Holling. *Panarchy: Understanding Transformations in Human and Natural Systems.* Island Press, Washington DC, December 2001.

16. B. Herzberg, D. Bekerman and I. Zeifman. Breaking Down Mirai: An IoT DDoS Botnet Analysis, 2016.

17. R. Holden, D. V. Val, R. Burkhard and Sarah Nodwell: A network flow model for interdependent infrastructures at the local scale. *Saf. Sci.,* 53(Suppl. C):51–60, 2013.

18. C.S. Holling. Resilience and stability of ecological systems. *Annu. Rev. Ecol. Syst.,* 4(1):1–24, 1973.

19. White House. *EO 13010 Critical Infrastructures Protection,* 1996.

20. M. Hromada, L. Lukas, M. Matejdes, J. Valouch,L. Necesal, R. Richter and F. Kovarik. *System and Method of Assessing Critical Infrastructure Resilience.* SPBI, Ostrava, 2013.

21. ISO. ISO 31000:2009 Risk management—Principles and guidelines.

22. M. Keck and P. Sakdapolrak. What is social resilience? Lessons learned and ways forward. *Erkunde*, 67(1):5–19, 2013.

23. B. Krebs. KrebsOnSecurity Hit With Record DDoS, 2016.

24. L. Labaka, J. Hernantes and J. M. Sarriegi. A framework to improve the resilience of critical infrastructures. *Int. J. Disaster Resil. Built Environ.*, 6(4):409–423, 2015.

25. T. G. Lewis. *Network Science: Theory and Application*. Wiley, Hoboken, NJ, 2009.

26. T. G. Lewis. *Critical Infrastructure Protection in Homeland Security*. Wiley, Hoboken, NJ, 2 edition, 2015.

27. T. G. Lewis, T. J. Mackin and R. Darken. Critical infrastructure as complex emergent systems. *Int. J. Cyber Warfare Terror.*, 1(1):1–12, 2011.

28. T. Lovecek, E. Sventekova, L. Maris and D. Rehak. Determining the resilience of transport critical infrastructure element: Use case. *In Proceedings of 21st International Conference (Transport Means 2017)*, 21, 824–828, Kaunas University of Technology, Kaunas, Lithuania, 2017.

29. J. Markuci. *Determining the synergic effect of national critical infrastructure [thesis]*. VSB Technical University of Ostrava, Ostrava, Czech Republic, 2014.

30. National Research Council. *Network Science*. National Academies Press, Washington, DC, 2005.

31. NIAC. *Critical Infrastructure Resilience: Final Report and Recommendations*, 2009.

32. OECD. *Oslo Manual: Guidelines for Collecting and Interpreting Innovation Data*. OECD, Paris, 3 edition, 2005.

33. G. Oliva, S. Panzieri and R. Setola. Agent-based input output interdependency model. *Int. J. Crit. Infrastruct. Protect.*, 3(2):76–82, 2010.

34. S. Pacinda. Network analysis and KARS method. *Sci. Popul. Protect.*, 2(1):75–96, 2010.

35. PCCIP. *Critical Foundations: Protecting America's Infrastructures*. President's Commission on Critical Infrastructure Protection, Washington, DC, 1997.

36. P. Pederson, D. Dudenhoeffer, S. Hartley and M. Permann. *Critical Infrastructure Interdependency Modeling: A Survey of U. S. and International Research*. INL, Idaho Falls, ID, 2006.

37. C. Perrow. *Normal Accidents: Living with High-Risk Technologies*. Princeton University Press, Princeton, NJ, 2 edition, 1999.

38. Public Safety Canada. *Action Plan for Critical Infrastructure (2014-2017)*. Public Safety Canada, Ottawa, 2014.

39. D. Rehak, J. Markuci, M. Hromada and K. Barcova. Quantitative evaluation of the synergistic effects of failures in a critical infrastructure system. *Int. J. Crit. Infrastruct. Protect.*, 14:3–17, 2016.

40. T. Rid. *Cyber War Will Not Take Place.* C Hurst & Co Publishers Ltd, London, 2013.

41. J. Riha. Vulnerability of infrastructure and environmental systems. *Spektrum*, 8(1):22–27, 2008.

42. S.M. Rinaldi, J.P. Peerenboom and T.K. Kelly. Identifying, understanding, and analyzing critical infrastructure interdependencies. *IEEE Control Syst.*, 21(6):11–25, 2001.

43. M. Senovsky, V. Adamec and P. Senovsky. *Critical Infrastructure Protection.* SPBI, Ostrava, 2007.

44. R. Setola, S. De Porcellinis and M. Sforna. Critical infrastructure dependency assessment using the input output inoperability model. *Int. J. Crit. Infrastruct. Protect.*, 2(4):170–178, 2009.

45. F. R. Spellman and R. M. Bieber. *Energy Infrastructure Protection and Homeland Security.* Government Institutes, Plymouth, 2010.

46. P. Trucco, E. Cagno and M. De Ambroggi. Dynamic functional modelling of vulnerability and interoperability of critical infrastructures. *Reliab. Eng. Syst. Saf.*, 105 (Suppl. C):51–63, 2012.

47. UN. *2009 UNISDR Terminology on Disaster Risk Reduction.* OSN, Geneva, 2009.

48. J. Valasek and F. Kovarik. *Crisis Management in Non-Military Crisis Situations: Module C. Ministry of Interior—The General Directorate of Fire Rescue Service of CR*, Czech Republic, Prague, 2008.

49. B. Walker, C. S. Holling, S. Carpenter and A. Kinzig. Resilience, adaptability and transformability in social ecological systems. *Ecol. Soc.*, 9(2), 2004.

50. D. J. Watts and S. H. Strogatz. Collective dynamics of small-world networks. *Nature*, 393(6684):440–442, 1998.

51. K. Zetter. *Countdown to Zero Day: Stuxnet and the Launch of the World's First Digital Weapon.* Broadway Books, New York, NY, 2015.

# Chapter 7

# Evolving Networks and Their Vulnerabilities

Abbe Mowshowitz, Valia Mitsou, and Graham Bent

## CONTENTS

## 7.1  Introduction

Networks are typically built in the real world by connecting smaller networks together to form larger ones. For example, two or more individual machines connected on a hub or a switch form a PAN (Personal Area Network). A CAN (Campus Area Network) connects two or more LANs (Local Area Networks) and is limited to a specific and contiguous geographical area such as a college campus, industrial complex, or a military base. A MAN (Metropolitan Area Network) connects two or more LANs or CANs together and is limited to the geographical bounds of a city. An internetwork is created by connecting two or more networks or network segments using routers, and so on. These examples of "natural" network growth suggest modeling networks as combinations of graphs. Some graph combinations—binary operations—that appear in the literature are defined below. The discussion focuses on the diameter and vulnerability of network combinations.

The combinational approach provides a mechanism for controlling growth and for analyzing network properties. On the one hand, it is possible, in principle, to control growth so as to realize networks with low diameter or low vulnerability; on the other hand, it is possible to decompose a large network with respect to some binary operation for the purpose of determining its diameter and vulnerability. *Diameter* is used in the usual sense to denote the maximum distance between any two vertices in the graph, where distance between vertices $u$ and $v$ is the number of edges in a shortest path between $u$ and $v$. *Vulnerability* is given a new graph-theoretic formulation that takes account of determining a set of vertices or edges whose removal breaks the network into two or more parts disconnected from each other.

The chapter highlights several binary operations on graphs that appear in the literature and tabulates the diameter and the vertex and edge vulnerability of graphs resulting from these operations. Some of the computations appeared in Reference 1 and some are reported here for the first time. Tables 7.1, 7.2 and 7.3 summarize the results obtained thus far, that is, the diameter and the vertex and edge vulnerability of selected network combinations based on the diameters and vulnerability of their respective components.

The main innovation in this chapter is a definition of network vulnerability as a three-component vector. To the best of our knowledge, most of the graph-theoretic definitions for vertex and edge vulnerability that appear in the literature consider the size of the set of vertices (or edges) that must be removed in order to disconnect the graph under various conditions. These conditions specify that the graph be broken into roughly equal components (minimum balanced separator [11]), or that the number of connected components of the "broken" graph be large (vertex integrity [3]), or that the size of the largest component of the "broken" graph be small (scattering number [9]). By contrast, the definition proposed here takes account of three aspects of network vulnerability, namely, the size of the vulnerability set, the entropy of the remaining graph, and the time needed for the adversary to determine the vulnerability set, thus defining a partial ordering among graphs. This refined concept allows for discriminating between cases of alleged vulnerability that

**TABLE 7.1:**    Diameter of a combined graph

| Operation | Vertices | Edges | Diameter |
|---|---|---|---|
| Union | $n_1 + n_2$ | $l_1 + l_2$ | $d = \infty$ |
| Join | $n_1 + n_2$ | $l_1 + l_2 + n_1 n_2$ | $d = min\{2, max\{d_1, d_2\}\}$ |
| Coalescence | $n_1 + n_2 - 1$ | $l_1 + l_2$ | $max\{d_1, d_2\} \le d$ $\le d_1 + d_2$ |
| Cartesian product | $n_1 n_2$ | $n_1 l_2 + n_2 l_1$ | $d = d_1 + d_2$ |
| Composition | $n_1 n_2$ | $l_2 n_1 + l_1 n_2^2$ | $d = max\{d_1, min\{d_2, 2\}\}$ |
| Tensor | $n_1 n_2$ | $2 l_1 l_2$ | $d \ge max\{d_1, d_2\}$ |
| Normal product | $n_1 n_2$ | $n_1 l_2 + n_2 l_1 + 2 l_1 l_2$ | $d = max\{d_1, d_2\}$ |
| Co-normal product | $n_1 n_2$ | $l_2 n_1^2 + l_1 n_2^2 - l_1 l_2$ | $d = min\{d_1, d_2\}$ |
| Rooted product | $n_1 n_2$ | $l_1 + n_1 l_2$ | $d \le d_1 + 2 d_2$ |
| Compounding | $n_1 n_2$ | $l_1 + n_1 l_2$ | $d \le d_2(d_1 + 1)$ |
| Star product | $n_1 n_2$ | $n_1 l_2 + n_2 l_1$ | $d \le d_1 + d_2$ |

**TABLE 7.2:** Vertex vulnerability of combined graph

| Product | Number edges | Vertex vuln. | Entropy (lb) | Time |
|---|---|---|---|---|
| Cartesian | $2nl$ | $cn$ | $b$ | $t$ |
| Star | $2nl$ | $cn$ | $b$ | $t$ |
| Composition | $ln(n+1)$ | $cn$ | $b$ | $t$ |
| Tensor | $2l^2$ | $cn$ | $b$ | $t$ |
| Normal | $2l(n+l)$ | $cn$ | $b$ | $t$ |
| Co-normal | $2l(n^2-l)$ | n/a | n/a | n/a |
| Compounding | $l(n+1)$ | $cn$ | $b$ | $t$ |
| Rooted | $l(n+1)$ | $c$ | $\to b$ | $t+n$ |

**TABLE 7.3:** Vulnerability of combined graph

| Product | Number edges | Vertex vuln. | Entropy (lb) | Time |
|---|---|---|---|---|
| Cartesian | $2nl$ | $cn$ | $b$ | $t$ |
| Star | $2nl$ | $cn$ | $b$ | $t+cn^2$ |
| Composition | $ln(n+1)$ | $n^2c$ | $b$ | $t$ |
| Tensor | $2l^2$ | $2c^2$ | $b$ | $t$ |
| Normal | $2l(n+l)$ | $cn+2c^2$ | $b$ | $t$ |
| Co-normal | $2l(n^2-l)$ | n/a | n/a | n/a |
| Compounding | $l(n+1)$ | $c$ | $b$ | $t+cn^2$ |
| Rooted | $l(n+1)$ | $c$ | $b$ | $t+n$ |

are not captured by existing definitions. Consider two graphs, one having a small vulnerability set and one having a larger one which can be found in much less time. It is not clear which of them should be considered as the most vulnerable so these two graphs probably should not be comparable on a vulnerability scale. Further implications of the proposed definition are examined in Section 7.4.

Section 7.2 presents the definitions of some basic graph operations that can be used for creating a larger graph from two smaller ones. Section 7.3 examines several properties of the resulting graphs, namely, the number of vertices and edges, and the diameter. Section 7.4 focuses on vertex and edge vulnerability, showing the vulnerability of graphs obtained by each of the graph operations presented in Section 7.2 in terms of the vulnerability of the constituent graphs. Section 7.5 presents an overview of the findings and indicates some key open problems.

## 7.2 Binary Operations on Graphs

Binary operations on graphs have proven useful in many problem areas. The extensive literature on product-based architectures such as hypercubes, meshes of trees, butterflies, and cube connected cycles document this claim (see, e.g., [8]). Simple operations and products are also useful in analyzing properties of sensor networks. In particular, we show how network diameter and vulnerability can be determined using such operations on graphs. The operations defined below include many of the more prominent ones discussed in the literature.

## 7.2.1   Simple operations

*(Disjoint) Union.* The (disjoint) union of two graphs $H_1(V_1,E_1)$, $H_2(V_2,E_2)$ is a graph $G(V,E)$, where $V = V_1 \cup V_2$ and $E = E_1 \cup E_2$.

*Join.* The join of two graphs $H_1(V_1,E_1)$, $H_2(V_2,E_2)$ is a graph $G(V,E)$, where $V = V_1 \cup V_2$ and $E = E_1 \cup E_2 \cup \{(x,y)|x \in V_1 \ \& \ y \in V_2\}$.

*Partial Join.* A partial join of two graphs $H_1(V_1,E_1)$, $H_2(V_2,E_2)$ is a graph $G(V,E)$, where $V = V_1 \cup V_2$ and $E = E_1 \cup E_2 \cup F$, where $F Z V_1 K V_2$.

*Skew Join.* A skew join of two graphs $H_1(V_1,E_1)$, $H_2(V_2,E_2)$ is a graph $G(V,E)$, where $V = V_1 \cup V_2$ and $E = E_1 \cup E_2 \cup \{(x,y)|x \in V_1' \ (V_1'$ is a subset of $V_1) \ \& \ y \in V_2\}$. This graph operation is defined in Reference 15

*Coalescence.* A coalescence of two graphs $H_1$ and $H_2$ is any graph $G$ obtained from the disjoint union of $H_1$ and $H_2$ by merging one vertex from $H_1$ and one from $H_2$. This graph operation is defined in Reference 2

## 7.2.2   Graph products

*Cartesian Product.* The cartesian product of two graphs $H_1(V_1,E_1)$ and $H_2(V_2,E_2)$ is a graph $G(V,E)$ with $V = V_1 \times V_2$ and $E = \{(u_1,u_2)(v_1,v_2)$, where $(u_1 = v_1 \ \& \ u_2v_2 \in E_2)$ or $(u_2 = v_2 \ \& \ u_1v_1 \in E_1)\}$.

*Composition (or Lexicographic) Product.* The composition product of two graphs $H_1(V_1,E_1)$ and $H_2(V_2,E_2)$ is a graph $G(V,E)$ with $V = V_1 \times V_2$ and $E = \{(u_1,u_2)(v_1,v_2)$, where $(u_1v_1 \in E_1)$ or $(u_1 = v_1 \ \& \ u_2v_2 \in E_2)\}$.

*Tensor Product (Direct Product, Categorical Product, Cardinal Product, or Kronecker Product).* The tensor product of two graphs $H_1(V_1,E_1)$ and $H_2(V_2,E_2)$ is a graph $G(V,E)$ with $V = V_1 \times V_2$ and $E = \{(u_1,u_2)(v_1,v_2)$, where $u_1v_1 \in E_1$ and $u_2v_2 \in E_2\}$.

*Normal Product (or Strong Product or And Product).* The Normal product of two graphs $H_1(V_1,E_1)$ and $H_2(V_2,E_2)$ is a graph $G(V,E)$ with $V = V_1 \times V_2$ and $E = \{(u_1,u_2)(v_1,v_2)\}$, where $(u_1 = v_1 \ \& \ u_2v_2 \in E_2)$ or $(u_2 = v_2 \ \& \ u_1v_1 \in E_1)$ or $(u_1v_1 \in E_1 \ \& \ u_2v_2 \in E_2)\}$.

*Co-normal Product (Disjunctive Product or Or Product).* The co-normal product of two graphs $H_1(V_1,E_1)$ and $H_2(V_2,E_2)$ is a graph $G(V,E)$ with $V = V_1 \times V_2$ and $E = \{(u_1,u_2)(v_1,v_2)$, where $u_1v_1 \in E_1$ or $u_2v_2 \in E_2\}$.

*Compounding of Graphs.* The compounding of two graphs $H_1(V_1,E_1)$ and $H_2$ is obtained by taking $|V_1|$ copies of $H_2$, indexed by the vertices of $H_1$, and joining two arbitrary nodes of copies $H_u$, $H_v$ of $H_2$ by a single edge whenever $uv$ is an edge in $E_1$ [7].

*Rooted Product.* The rooted product of a graph $H_1(V_1, E_1)$ and a rooted graph $H_2$ is a subgraph of the Cartesian product consisting of $|V_1|$ copies of $H_2$ together with edges joining the root nodes of the $i$-th and $j$-th copies if vertices $u_i$ and $u_j$ are adjacent in $H_1$. The rooted product is a special case of the compounding product [6].

*Star Product.* The Star product of two graphs $H_1(V_1, E_1)$ and $H_2(V_2, E_2)$ consists of $|V_1|$ copies of $H_2$ indexed by the vertices of $H_1$, together with edges between the copies determined as follows: For every edge $e = uv$ in $E_1$ define a bijection $f_e : V_2 \to V_2$. Edges joining corresponding copies $H_u$ and $H_v$ of $H_2$ are determined by the function. In particular, each vertex $x$ in copy $H_u$ of $H_2$ is adjacent to vertex $f_e(x)$ of the $H_v$. This means that for every edge in $H_1$ the corresponding copies of $H_2$ are joined by matching their respective vertices [5].

## 7.3 Network Diameter

The operations defined in the previous section illustrate how new networks can be formed by combining existing ones. Now we show how these combinations can be analyzed so as to obtain information about their properties in terms of corresponding properties of the constituent networks. Thus, we obtain the number of vertices, the number of edges and the diameter of the resulting graph for each of the binary operations shown above in terms of the corresponding characteristics of the constituent graphs. The information is collected in Table 7.1.

For the reminder of the section the notation $s \to t$ denotes a shortest path in a graph from vertex $s$ to vertex $t$. Thus, the diameter of a graph is given by max $\{s \to t\}$, where max is taken over all possible pairs of vertices $s$ and $t$ in the graph. In what follows, $G$ denotes the resulting graph of an operation on graphs $H_1$ and $H_2$ which have $n_1$ and $n_2$ vertices, $l_1$ and $l_2$ edges and diameters $d_1$ and $d_2$, respectively.

The diameters of the *union*, *join*, and *coalescence* are immediate consequences of their respective definitions.

*Cartesian Product*: The cartesian product is formed by substituting a copy of $H_2$ for each vertex of $H_1$ and by joining corresponding vertices of different copies for each edge of $H_1$. This means vertices $(u_1, u_2)$ and $(v_1, u_2)$ are adjacent in $G$ when $u_1 v_1$ is an edge in $H_1$. Thus the total number of edges is $n_1 l_2$ ($n_1$ copies of $H_2$ where each one has $l_2$ edges) plus $n_2 l_1$ ($l_1$ groups of edges between the copies and each group has exactly $n_2$ edges, one for every node in the each copy of $H_2$). A path from vertex $s = (s_1, s_2)$ to $t = (t_1, t_2)$ can be obtained combining a path $(s_1, s_2) \to (t_1, s_2)$ with a path $(t_1, s_2) \to (t_1, t_2)$. Observe that there is no shorter path between $s$ and $t$ so their distance is equal to the sum of the distances of $s_1 \to t_1$ and $s_2 \to t_2$. The diameter of $G$ is thus $d_1 + d_2$.

*Composition*: The composition is formed by replacing each vertex of $H_1$ with a copy of $H_2$ and joining all the vertices of two copies whenever the corresponding vertices of $H_1$ are adjacent. The total number of edges is $n_1 l_2$ ($n_1$ copies of $H_2$

where each one has $l_2$ edges) plus $n_2^2 l_1$ ($l_1$ groups of edges between the copies, each group having exactly $n_2^2$ edges, one for every possible pair of nodes between two copies of $H_2$ that correspond to adjacent vertices in $H_1$). If $s$ and $t$ are in the same copy of $H_2$ ($s = (s_1, s_2)$ and $t = (s_1, t_2)$) then the shortest path from $s$ to $t$ passes through the vertex $(u_1, t_2)$—where $(s_1, u_1)$ is an edge in $H_1$—making the distance between $s$ and $t$ equal to 2 unless $(s_1, s_2)$ and $(s_1, t_2)$ are adjacent (where $s_2 t_2$ is an edge in $H_2$). Thus, the maximum distance between any two vertices of the same copy is 2 unless $H_2$ is a clique where $d_2$ is smaller than 2. If $s = (s_1, s_2)$ and $t = (t_1, t_2)$ belong to different copies, then the shortest path between them passes through the vertices $(u_i, t_2)$, where $u_i$ denotes a vertex included in the path $s_1 \rightarrow t_1$. Thus, the distance between $s$ and $t$ is equal to the distance between $s_1$ and $t_1$. The maximum distance between every pair of vertices in different copies is equal to $d_1$. Thus the diameter of $G$ will be the maximum of the first and the second cases.

*Tensor Product*: Let $A_1$ and $A_2$ be the adjacency matrices of $H_1$ and $H_2$, respectively. The adjacency matrix $A$ of $G$ can be formed as follows: for each 0 entry in $A_1$, substitute the 0 matrix, and for each 1 of $A_1$, substitute the matrix $A_2$. Characteristics of $G$ can be read from $A$. Since $A_1$ contains $2l_1$ ones and every one represents $A_2$—containing $2l_2$ ones—there are $4l_1 l_2$ ones in $A$. Thus, $G$ has $2l_1 l_2$ edges. $G$ has an edge joining vertices $(u_1, u_2)$ and $(v_1, v_2)$ if both $u_1 v_1$ and $u_2 v_2$ are edges in $H_1$ and $H_2$, respectively. Thus, the length of a shortest path from $s = (s_1, s_2)$ to $t = (t_1, t_2)$ is at least as great as a shortest path from $s_1$ to $t_1$ in $H_1$ and a shortest path from $s_2$ to $t_2$ in $H_2$, giving the diameter shown in Table 7.1. Note that $G$ might be disconnected.

*Normal Product*: The normal product is a combination of the Cartesian and the tensor product. The total number of edges of $G$ in this case is the sum of the edges of the Cartesian and the tensor product. Since the graphs are loop-free, no edge is counted twice. A shortest path from $s = (s_1, s_2)$ to $t = (t_1, t_2)$ consists of a shortest path $s_1 \rightarrow t_1$ together with a shortest path $s_2 \rightarrow t_2$. If $s_1 \rightarrow t_1$ is shorter than $s_2 \rightarrow t_2$, then the path contains all the vertices $(t_1, u_i)$ of $G$, where $u_i$ denotes a vertex included in $s_2 \rightarrow t_2$. Thus, $s \rightarrow t$ has the same length as $s_2 \rightarrow t_2$. Similarly if $s_1 \rightarrow t_1$ is longer than $s_2 \rightarrow t_2$, the path $s \rightarrow t$ has the same length as $s_1 \rightarrow t_1$, so the diameter of $G$ equals $\max\{d_1, d_2\}$.

*Co-normal Product*: The co-normal product graph $G$ is obtained by taking $n_1$ copies of $H_2$ and joining each pair of vertices belonging to copies that correspond to connected vertices of $H_2$. This contributes $n_2^2 l_1$ edges. Since the operation is symmetric, $G$ can be obtained by taking $n_2$ copies of $H_1$ and connecting pairs of different copies in the same way as above, resulting in an additional $n_1^2 l_2$ edges.

However, some edges have been included twice in the foregoing account: if $u_1 v_1$ and $u_2 v_2$ are edges of $H_1$ and $H_2$, respectively, then the edge between the nodes $(u_1, u_2)$ and $(v_1, v_2)$ is included in both steps: one time in counting edges between copies of $H_2$, where $u_1 v_1$ is an edge of $H_1$ and one time in counting edges between copies of $H_1$ because $u_2 v_2$ is an edge of $H_2$. The number of edges counted in the sum is $2l_1 l_2$ because if edges $u_1 v_1$ and $u_2 v_2$ are both included this means edges $u_1u_2$, $v_1v_2$, and $(u_2,u_1)(v_2,v_1)$ are included as well and only half of them should be counted. Thus, the total number of edges is $n_1^2 l_2 + n_2^2 l_1 - l_1 l_2$. To obtain the diameter, note that a shortest path from $s = (s_1,s_2)$ to $t = (t_1,t_2)$ consists either of edges of the form $(u_1,w)(v_1,w)$, where $u_1v_1$ is an edge of $H_1$, or those of the form $(w,u_2)(w,v_2)$, where $u_2v_2$ is an edge of $H_2$. Thus, if $s_1 \rightarrow t_1$ is shorter than $s_2 \rightarrow t_2$, a shortest path between $s$ and $t$ includes all the edges of the form $(s_1,s_2)(u_1,t_2)$, where $u_1$ denotes a vertex in the path $s_1 \rightarrow t_1$. The analogous argument applies if $s_2 \rightarrow t_2$ is shorter than $s_1 \rightarrow t_1$. Thus, the diameter is the min of $d_1$ and $d_2$.

*Star Product*: The star product of two graphs $H_1$ and $H_2$ is a generalization of the Cartesian product. Every vertex of a copy $H_u$ of $H_2$ is connected to exactly one vertex of the copy $H_v$ whenever $(u,v)$ is an edge in $H_1$. Any path from vertex $s$ (in $H_u$) to vertex $t$ (in $H_v$) must pass through the intermediate copies of $H_u$ and $H_v$, covering that way distance at most $d_1$ (if the distance between $u$ and $v$ in $H_1$ is $d_1$). The worst case occurs when it is necessary to traverse distance $d_2$ inside the copies (as in the Cartesian product). The diameter then becomes $d_1 + d_2$.

---

## 7.4   Network Vulnerability

In this section, we analyze the vulnerability to network attacks formed from the binary product operations discussed above. For simplicity of presentation, we will assume that a network (denoted by $G$) is the product of graph $H$ with itself. An attack on a network $H$ is represented by the removal of a set $X$ of vertices or edges chosen by an adversary for the purpose of disconnecting $H$. The graph resulting from the removal of $X$ from $H$ is denoted by $R$. The (connected) components of $R$ are denoted by $R_1, R_2, \dots, R_k$.

Vulnerability depends on the knowledge and resources available to an attacker as well as the structure of the network itself. It also depends on the time required to destroy a network before defensive measures can be taken to ward off an attack.

*Knowledge* refers to what a potential attacker knows about the structure of the network. It appears that some network structures are more vulnerable to attack than others. But it is far from trivial to determine the structure of a network if one has no *a priori* knowledge of how it was built. Clearly, the more a potential attacker knows about the structure of a network or about the protocol governing its growth, the greater the likelihood of mounting a successful attack.

A proxy for *resources* is the cost of mounting an attack. Note that a network that could, in principle, be destroyed by an undertaking that no one could afford is not vulnerable to such an undertaking. A complete specification of the resources would include details of the computational power needed to determine the elements to be attacked, and the physical actions required to destroy the actual network nodes.

A network that can be broken easily into several isolated pieces may be vulnerable to attack. Note that if the vast majority of the nodes are concentrated in one large connected component, an attack cannot be considered successful. Furthermore, assuming that an attack disconnects the network into components of comparable sizes, the more the components the greater the damage. This observation provides a useful basis for a definition of vulnerability.

The makeup (i.e., number and sizes of the components) of a graph $R$ resulting from removal of elements can be measured by the entropy of a vertex partition. The vertex sets of the components of $R$ are pairwise disjoint and their union contains $m$ vertices. Thus a finite probability scheme, associated with the $k$ components $R_i$ of $R$, can be defined as follows: $p_i = |R_i|/m$ $(1 \leq i \leq k)$. The entropy of the finite probability scheme $-\sum_{i=1}^{k} p_i \log p_i$ provides a measure of the number and relative sizes of the components (see [6] for a survey of applications of entropy in graph theory).

In the case of two almost equal connected components, the partition has entropy close to log 2. If there are $O(n)$ components or isolated vertices, the partition has entropy close to log $n$. Furthermore, if just a small number of vertices become isolated and the rest of the graph remains connected, the partition has entropy close to 0.

Previous attempts to define vulnerability, like the minimum balanced separator, the vertex integrity, and the scattering number consider limited aspects of the structure of the remaining graph. Some definitions insist the remaining graph be disconnected into roughly equal components; others require that the original graph be broken into a certain number of components; yet others place constraints on the size of the largest component. They all fail to capture the whole structure of the remaining graph. Graph entropy provides a more complete measure of the structure of the remaining graph in the sense that it takes account of both the number of connected components that $R$ contains and their respective sizes.

A network, modeled as a graph with $n$ vertices, could, in principle, be broken into pieces by the removal of elements (vertices or edges). Let $X$ be a set of elements of a graph $H$ whose removal results in a graph $R$ consisting of $k$ ($>1$) non-null components $R_1, R_2, \ldots, R_k$. Let $I$ be the entropy of the partition associated with the $R_i$, and let $\delta > 0$ be a (threshold) value specified by the network owner or manager. If $I > \delta$, then $X$ is a *vulnerability set* of $H$. Note that the complete graph $K_n$ does not have a vulnerability set for any $\delta$, since $n - 1$ vertices must be removed to disconnect $K_n$, thus leaving a partition with zero entropy. Graphs lacking a vulnerability set are clearly not vulnerable. By introducing the parameter $\delta$, we allow our definition some flexibility as the concept of vulnerability may vary significantly with different applications.

To arrive at a precise formulation the ambiguities in the statement must be resolved. In particular, one must characterize the means by which a network can be broken up. In addition, bounds on the complexity of the algorithm needed to compute a vulnerability set must be stated.

There are two very different kinds of operations undertaken by an attacker, namely, physical and logical. The physical operation is the effort an attacker must expend in eliminating the actual network nodes or edges identified in a vulnerability set. Malicious attacks are characterized by removal of specific vertices or edges or both vertices and edges. Some of the methods that have been used to remove network elements include destruction of hardware, infection of software with viruses, denial of service attacks, and so on. The aim of all these methods is to break the network into pieces that cannot communicate with each other.

Possible measures of the physical operation include the cost and the amount of time required to mount a successful attack. The latter measure should be a slowly growing function of the size of the network. An attack requiring $O(n^2)$ seconds, say, to destroy a network with $n = 10^6$ nodes is not likely to succeed. Although it is an important issue, the physical operation is beyond the scope of this chapter.

We are mainly concerned here with the logical operation of identifying the vulnerability set. Three aspects of this operation need to be taken into account: (a) the size of the vulnerability set, (b) the entropy of the partition resulting from its removal, and (c) the computational effort required to identify the elements of the vulnerability set. From the foregoing, it is clear that vulnerability cannot be measured as a scalar quantity. Leaving aside the physical effort associated with an attack, a *vulnerability vector* for a graph can be specified as a three-component vector $v = (c, b, t)$, where $c$ is the size of a vulnerability set, $b$ is the entropy of the partition associated with the graph resulting from the removal of a vulnerability set, and $t$ is the complexity of the algorithm that computes the vulnerability set.

As defined here, a vulnerability vector is not unique. A graph may have vulnerability sets of varying sizes, and it is possible that a larger vulnerability set could be determined more quickly than a smaller one. Moreover, two vulnerability sets of the same order could yield partitions with different entropy. Although two networks may not be comparable with respect to the vulnerability vector, this measure does allow for constructing a partial order of networks. If two networks agree on the first two components of the vulnerability vector, they can be distinguished according to the growth rate of the third component. For example, the algorithm for determining a vulnerability set may grow as $\log n$ or as $\log \log n$.

In what follows, we analyze the vulnerability of the network products defined above. Several efforts to compute the vulnerability of products of graphs and recursive networks already appear in the literature [10,12,16]. However, our perspective is different in that we examine a large variety of graph product operations that could generally serve as a way to model network growth. For each product, we determine two vulnerability vectors, one based on a vulnerability set of vertices and the other on a set of edges.

The values in Tables 7.1 and 7.2 assume the network graph $G(V,E)$ is the result of a binary operation on two graphs $H_1(V_1, E_1)$ and $H_2(V_2, E_2)$. For simplicity, both $H_1$ and $H_2$ are taken to be isomorphic to the same graph $H$ with vertex (edge) vulnerability set of size $c$.

**Remark 7.1.** *The size of a vertex vulnerability set is less than or equal to the size of an edge vulnerability set. This follows from the fact that the vertex connectivity of a graph is less than or equal to its edge connectivity.*

**Remark 7.2.** *If a graph $H$ is disconnected into two subgraphs $R_1$ and $R_2$ by the removal of a vulnerability set, then the resulting partition has entropy at least $-(p_1 \cdot \log p_1 + p_2 \cdot \log p_2)$, where $p_i = m_i/m$ with $m_1$ and $m_2$ the number of nodes of $R_1$ and $R_2$, respectively, and $m_1 + m_2 = m$. This is easily verified by computing the entropy of the partition.*

In what follows we assume that $H$ (consisting of $n$ nodes) has a vulnerability vector $(c, b, t)$ (see Figure 7.1) and obtain the graph $G$ by taking one of the aforementioned products of $H$ with itself. The adversary has identified the set $X$ of $c$ vertices or edges of $H$ whose removal would result in a graph $R$ with $k$ components $R_1, R_2, \ldots, R_k$ of sizes $m_1, m_2, \ldots, m_k$, respectively, whose associated vertex partition has entropy $b > \delta$ (the threshold value for the network). The adversary wants to disconnect the graph $G$. However, computing a vulnerability set of $G$ from scratch would require much computational effort and would probably be difficult or even impossible since $G$ is much larger than $H$. A good strategy for the adversary would be to use the information that is known about disconnecting $H$ in attacking the product $G$. In this case, the adversary intends to disconnect $G$ by disconnecting those parts $P_1, P_2, \ldots, P_m$ of $G$ that correspond to $R_1, R_2, \ldots, R_m$ (each of the $P_i$s contains $k_i$ sets of $n$ nodes which are connected according to the rules specified by the product). This will create at least $m$ (not necessarily) connected parts depending on the particular product operation involved. The vulnerability vector $(c', b', t')$ of $G$ is to be expressed in terms of the components $c$, $b$, and $t$ of the vulnerability vector of $H$.

Note that the product graphs in Tables 7.2 and 7.3 all have $n^2$ vertices.

### 7.4.1   Vertex vulnerability

All the product operations examined here, save the co-normal product, have an easily determined upper bound on the first component of the vulnerability vector. This can be seen as follows. Remove from $G$ all those copies of $H$ that correspond to nodes of the vulnerability set $X$ of $H$. Now, observe that the set

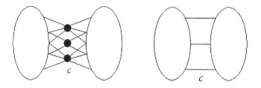

**FIGURE 7.1:** The left figure represents the vertex vulnerability set and the right one the edge vulnerability set of graph $H$.

of edges of the composition product is a superset of the set of edges of any product operation considered here other than the co-normal product, since each of them only adds edges between copies of $H$ that correspond to neighboring vertices of $H$ and the composition product adds all the possible edges between any two neighboring copies. So, a vulnerability set for the composition product is also a vulnerability set for any other of the remaining products in question. Let $n$ be the number of vertices of $H$, and $c$ be the number of vertices in the vulnerability set $X$ of $H$ (i.e., the set of vertices whose removal disconnects $H$). It is obvious that removing $cn$ vertices (from those copies of $H$ that correspond to the $X$ set of $H$) would disconnect the composition product graph $G = H \times H$.

On the other hand, this seems to be best possible if the adversary wants to rely on knowledge of $X$ without looking for a new vulnerability set in $G$. For example, one other approach would be to consider removing the $c^2$ vertices $(u,v)$, where both $u$ and $v$ are in $X$. However, the removal of such a set would not necessarily result in a disconnected graph.

The rooted product is the most vulnerable. Removing just $c$ vertices can disconnect $G$, namely the roots of the copies that correspond to the vertices of the vulnerability set $X$. Note that this does not occur in the case of the compounding product because the addition of a random factor (namely, the vertices that are going to be connected in each two copies) decreases the vulnerability of the network. The definition of the compounding product allows for joining different vertices for each edge in $H$ and thus all the $n$ vertices of the copies corresponding to the vertices of the vulnerability set would have to be eliminated in order to ensure that $G$ is disconnected.

The co-normal product is a special case. Recall the construction of the co-normal product of a graph $H(V,E)$ with itself. For any two vertices $w_1$, $w_2$ in $V$ and any edge $(u,v)$ in $E$, all the edges $((u,w_1),(v,w_2)$ and $((w_1,u),(w_2,v))$ are included. This product has edges from any $P_i$ to any $P_j$ (see, e.g., Figure 7.2), so if we only remove the copies corresponding to elements in the set $X$, the graph

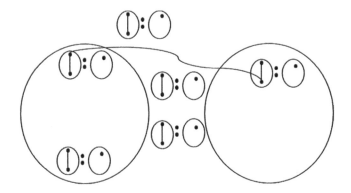

**FIGURE 7.2:** An edge going from $P_1$ to $P_2$ when $G$ is the co-normal product. Knowledge of the vulnerability set $X$ of $H$ is useless.

will not be disconnected. Furthermore, the graph will not be disconnected by removing the elements of $X$ sets from all the copies of $H$ in $G$. So, the fact that $H$ is vulnerable by itself does not suggest a method of attack to an adversary. Additional information about every edge inside the $R_i$s is required to construct a vulnerability set of $G$.

Turning to the second component (entropy) of the vulnerability vector, we examine the relationship between the entropy $b$ of $R$ and the entropy $b'$ of $P$, where $P$ is the subgraph of $G$ that corresponds to $R$, and $G$ is any of the above-mentioned products except for the co-normal (shown earlier to be invulnerable) and the rooted (discussed separately in the following paragraph). Note that $b'$ has to be at least $b$. If each of the $P_i$s are connected components, then

$$b' = -\sum_i \frac{|P_i|}{n^2 - cn} \log \frac{|P_i|}{n^2 - cn} = -\sum_i \frac{n \cdot m_i}{n^2 - cn} \log \frac{n \cdot m_i}{n^2 - cn} = b. \quad \text{Furthermore,}$$

Remark 7.2 states that if some of the $P_i$s are disconnected into two or more connected pieces, then the entropy of $P$ must be at least as much as it would be if all the $P_i$s were connected subgraphs. Thus, $b' \geq b$.

For the rooted product, removing $c$ nodes (instead of $cn$) leaves the graph disconnected into $k$ large parts $P_1, P_2, \ldots, P_k$, each of size $m_i \cdot n$ for $i = 1, \ldots, k$ and $c$ smaller parts of size $n - 1$. In this case, $b'$ is given by

$$b' = -\sum_i \left( \frac{m_i \cdot n}{n^2 - c} \log \frac{m_i \cdot n}{n^2 - c} \right) - c \frac{n - 1}{n^2 - c} \log \frac{n - 1}{n^2 - c} \approx b, \text{ when } n \text{ tends to infinity}$$

and $c \ll n$.

For purposes of analyzing the third component (i.e., time) of the vulnerability vector, it is necessary to make some assumptions about the labeling and representation of the network. In the absence of specific knowledge about the way the vertices of the product network are labeled, an adversary could not use any information that is available about $H$ unless some computationally infeasible problem is solved, that is, unless a correspondence is established between the labels attached to the vertices of the actual network and those of a representation in which the labels are fully determined by the original graph $H$—note that this is an instance of the well-studied graph isomorphism problem which is not known to be solvable in polynomial time. However, it is not unreasonable to assume that if the product structure of the network is known, then the coordinates of every vertex are known as well. In other words, in most practical situations, when a network is formed as a product graph, the network's designer will not attempt to disguise the network structure by permuting vertex labels.

As observed earlier, the copies of $H$ in $G$ are identical as labeled graphs and furthermore the labels of the copies of $H$ in $G$ are determined by the corresponding nodes in $H$. So if an adversary computes a vulnerability set of $H$, that same set can be applied immediately to determine the copies of $H$ that should be removed from $G$ (in other words, the vulnerability set of $G$). So, the time needed by an adversary to disconnect the network $G$ if $G$ is any product $H \times H$ other than the co-normal or the rooted products is just the time needed to determine a vulnerability set of $H$.

As indicated for the rooted product, it is not necessary to remove all the $cn$ vertices from $G$ in order to disconnect it. $G$ can be disconnected by just removing the root from every copy that corresponds to a vertex in the vulnerability set of $H$. However, an adversary would also have to identify the root. This requires searching only the $n$ nodes of $H$, since the label of the root is the same for every copy of $H$ in $G$. So the time required is $t + n$: $t$ for determining the vulnerability set of $H$, and $n$ for determining the root.

## 7.4.2 Edge vulnerability

Given a product graph $G = H \times H$, where $\times$ denotes any of the product operations discussed in Section 7.2, we should like to determine a vulnerability set consisting of edges under the assumption that $H$ is edge-vulnerable with vulnerability vector $(c, b, t)$. Such a vulnerability is a collection of edges whose removal disconnects $G$ and gives rise to a partition with entropy greater or equal than a given threshold value $\delta > 0$.

As stated above, finding a vulnerability set in $G$ from scratch would be a difficult task since $G$ contains $n^2$ vertices. A reasonable aim of an adversary would be first to solve the easier problem of determining a vulnerability set in $H$ and then follow some strategy proposed by this solution in order to achieve the more difficult goal of disconnecting $G$. A straightforward idea would be to remove all the edges connecting different $P_i$s, that is, those pieces of $G$ that correspond to the $R_i$s (the connected components in which $H$ is disconnected after a malicious attack). To determine whether or not such an attempt would be successful we need to express the vulnerability vector of $G$ in terms of the components $c$, $b$, and $t$ of the vulnerability vector of $H$ for a given vulnerability set.

In order to determine a vulnerability set of $G$, given a vulnerability set $X$ of $H$, we examine those edges of $G$ that correspond in some sense to edges in $X$. To make the previous statement more precise the following definitions are needed:

**Definition 7.1.** *The F-set of $H$ (denoted by $F$) is the set of vertices $u$ which are incident to edges that belong to the vulnerability set $X$ of $H$. More precisely, $F = \{u \mid \exists v : (u, v) \in X\}$. Furthermore, $F = \cup_i F_i$, with $F_i = F \cap R_i$.*

**Definition 7.2.** *The T-set of $G$ (denoted by $T$) is the set of copies that correspond to vertices in the F-set of $H$, that is, $T = \{H_u \mid u \in F\}$. Furthermore, $T = \cup_i T_i$, with $T_i = F \cap P_i$.*

With the exception of the co-normal product, $G$ can be disconnected by disconnecting copies belonging in $T$, since edges are added only between copies that correspond to adjacent vertices. So a possible vulnerability set of $G$ is the set of all those edges connecting the copies in different $T_i$s. Henceforth, we denote this set of edges by $Y$ (see Figure 7.3).

*Cartesian and Star Products*: In the Cartesian product, each vertex $(u, w)$ of a copy $H_u$ is adjacent to the vertex $(v, w)$ of a copy $H_v$ whenever $u$ and $v$ are

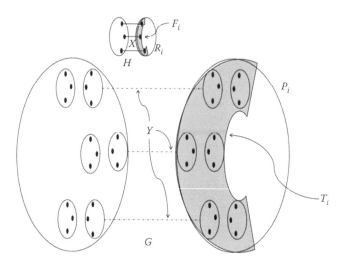

**FIGURE 7.3:** Copies of $H$ belonging in the $T$-set with internal vertices forming their $F$ sets.

neighboring vertices in $H$. Thus the Cartesian product creates a matching between each two copies $H_u$ and $H_v$ whenever $(u,v)$ is an edge in $H$. So, each edge in $X$ causes the placement of $n$ edges in $G$ (see Figure 7.4). Note that the set $Y = \{((u, w), (v, w))|(u, v) \in X\}$ is a vulnerability set of $G$, so $c' = |Y| = c \cdot n$.

The case of the Star product is similar: For each edge $(u,v)$ in $X$, the two copies $H_u$ and $H_v$ are connected with an arbitrary matching. Again, destroying the matching between copies belonging in different $T_i$s disconnects $G$. So $c' = c \cdot n$.

*Composition*: In the composition product, every edge of $H$ is associated with $n^2$ edges of $G$ since all vertices in adjacent copies of $H$ are adjacent. Once again,

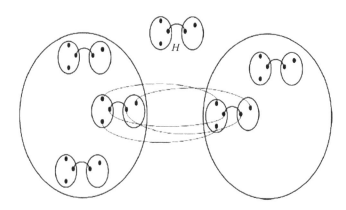

**FIGURE 7.4:** Edges added in the Cartesian product corresponding to a single edge in $X$.

disconnecting those copies that correspond to the $F$ sets would disconnect $G$ since no edge is placed between non-adjacent copies. Thus, $Y$ is a vulnerability set of $G$ and $c' = |Y| = cn^2$ edges must be removed to disconnect $G$.

*Tensor Product*: In the tensor product, there is an edge $(u_1, u_2)(v_1, v_2)$ in $G$ if and only if both $(u_1, v_1)$ and $(u_2, v_2)$ are edges of $H$. Thus, every edge in $X$ is associated with $2c$ edges in $Y$. So $2c^2$ edges must be removed to disconnect $G$ (see Figure 7.5). Note that, if $c$ is a constant number then removing only a constant amount of edges would disconnect $G$. This observation leads to the conclusion that the Tensor product of an edge-vulnerable graph $H$ by itself is also edge-vulnerable. In addition, observe that the tensor product of an edge-vulnerable graph by itself is also vertex-vulnerable due to Remark 7.1. By contrast, if $H$ is vertex-vulnerable, the size $c'$ of the vulnerability set of the tensor product $H \times H$ depends on both $n$ (the number of vertices of $H$) and $c$ (the size of the vulnerability set of $H$) as demonstrated in the previous section.

*Normal Product*: The normal product has the edges of both the Cartesian and the tensor product, and the sets of edges added by these products are disjoint. Thus $c' = cn + 2c^2$.

*Co-normal Product*: The co-normal product is the least vulnerable of all the products considered here. This product includes $n$ copies of $H$ in which every pair of vertices in adjacent copies is joined by an edge; in addition, the product includes edges of the form $((u_1, v_1), (u_2, v_2))$ for *every* pair $H_{u1}$, $H_{u2}$ of copies and every edge $(v_1, v_2)$ in $H$. As a result, corresponding to the edges of the $H_i$s, there exist edges joining vertices of different $P_i$s that do not belong in $Y$. Thus, disconnecting the $T_i$s is not sufficient to disconnect $G$ (see Figure 7.6). In addition to

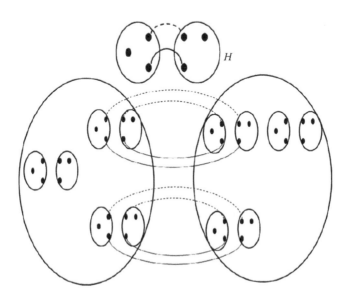

**FIGURE 7.5:** $X$ and $Y$ sets associated with the tensor product.

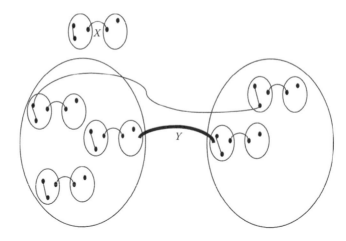

**FIGURE 7.6:** For the co-normal product $Y$ is not necessarily a vulnerability set.

the edges in $X$ an adversary must have information about every edge in $H$ in order to create a vulnerability set for $G$. For this reason, the co-normal product is deemed invulnerable.

*Rooted and Compounding Products*: Each pair of copies corresponding to adjacent vertices is joined by just one edge. Thus, the removal of the $c$ edges between the copies belonging in the $T$-set will disconnect $G$. Similar observations to those made for the Tensor product apply to the compounding product as well. The compounding product of an edge-vulnerable graph $H$ by itself is both vertex- and edge-vulnerable as opposed to the case of $H$ being vertex-vulnerable.

Having identified the vulnerability set $Y$ of $G = H \times H$ for every possible product operation $x$, we can now determine the entropy of the partition $P = \cup_i P_i$ that is created by the removal of $Y$. Computing the entropy component of the vulnerability vector of $G$ when $H$ is edge-vulnerable is straightforward. Since edges rather than vertices are being removed, the proportion of vertices in the $P_i$s is exactly the same as proportion of vertices in the $R_i$s. So for all but the co-normal product, the entropy component for $G$ is exactly the same as the entropy component for $H$ if the subgraphs induced by the $P_i$s are connected. Now, if some $P_i$s are disconnected then the entropy is at least as much as it would be if they were connected. So, $b' \geq b$.

Once an edge vulnerability set $X$ of $H$ has been determined, no additional effort is needed to specify the edges that should be removed from $G$, assuming the procedure for building $G$ from $H$ is deterministic and known to the adversary. So except for the co-normal which is known to be invulnerable and the compounding, rooted, and star products, an adversary needs time $t$ to determine the vulnerability set of $G$.

The compounding, rooted, and star products pose special challenges to an adversary. In these products, graph $R$ is not built deterministically, that is, the choice of vertices linking corresponding copies may differ from one application of the definition to another. Thus, it is necessary to examine these three products separately to determine the time required to determine a vulnerability set.

*Rooted Product*: If time $t$ is needed to determine those $c$ edges that disconnect $H$, additional effort is required to determine the set of edges that disconnect $G$. This additional effort entails determining the root of $H$ which can be accomplished by a one-time search among the $n$ nodes of some copy of $H$ adding at most $n$ to the overall time. So the adversary needs time $t + n$ to determine the vulnerability set of $G$.

*Compounding Product*: Each two neighboring copies in different $T_i$s are joined by exactly one edge. Removing this edge will disconnect the two copies. So, as in the preceding case, it is only necessary to remove $c$ edges to disconnect $G$. Note that randomness does not help to increase the size of the vulnerability set. However, it does help in providing a less vulnerable network since the time needed for the adversary to determine those $c$ edges in $G$ is no longer $t + n$. As indicated earlier, the adversary does not know which two vertices are joined, so in the worst case, for every one of the $c$ edges in a vulnerability set, all the $n \cdot n$ ($n$ for the first copy of $H$ and $n$ for the second copy of $H$) possible pairs of vertices must be examined. Thus, the time required to determine a vulnerability set of edges is $t + cn^2$ (Figure 7.7).

*Star Product*: The star product is very similar to the Cartesian product in terms of the size of the vulnerability set. However, it is much more difficult to determine $Y$ in the star product than in the Cartesian. For every edge in the vulnerability set $Y$, the adversary has to do the following. For every one of the $n$ vertices in the first copy, it is necessary to find the vertex in the second copy

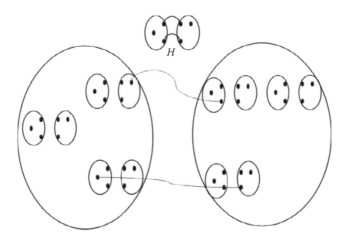

**FIGURE 7.7:** In the compounding product, the adversary does not know which two vertices of adjacent copies are joined.

to which it is adjacent. The time required to do this is $n^2$. This gives a total of $t + cn^2$, where $t$ represents the time to determine the vulnerability set of $H$ and $cn^2$ represents the time to determine the edges that connect each two copies corresponding to the vertices in the $F$-set of $H$.

The time $(cn^2)$ needed to find the edges that link the copies corresponding to the $F$ sets derives from a worst case analysis. However, an average case analysis does not significantly alter the result. The average time needed in the star product is always $cn^2$ since for every one of the $c$ edges in the vulnerability set of $H$ it is necessary to determine each one of the $n$ pairs of edges that connect the two corresponding copies of $H$. The average time needed in the compounding becomes $cn^2/4$, since for every one of the $c$ edges in the vulnerability set of $H$ it is necessary to determine the two vertices that are adjacent in the corresponding copies, a search that requires time $n/2$ on average for each of the two joined copies.

A comparison of the star and compounding products is quite revealing. In the star product, an adversary has to remove $cn$ edges, whereas only $c$ edges have to be removed in the compounding product. However, the time required to determine the $c$ edges of the vulnerability set for the compounding product is exactly the same as the time required to determine the $cn$ edges of the star product's vulnerability set. This highlights our contention that the size of a vulnerability set is not sufficient to measure a network's vulnerability.

---

## 7.5    Discussion and Conclusions

The point of departure for the work reported here is a novel interpretation of a sensor network. Such a network can be modeled as a distributed database by treating it as a dynamic distributed federated database system in which each sensor stores its data in its local database [4,14]. In such a configuration—called a Gaian Database—the sensor databases are interconnected in a way that makes the data from all databases accessible everywhere in the network within a certain number of hops.

Since the configuration of a Gaian Database is determined, in part, by the protocol for growth, questions about the implications of the protocol for the diameter and vulnerability of the evolving network arise quite naturally. One possibility that has been considered is connecting new nodes to others in the database according to the principle of preferential attachment. Another way—explored in this chapter—is to combine partially formed Gaian databases according to some graph operation. Minimization of diameter and vulnerability are natural criteria in the selection of a graph operation.

In Reference 14, we examined properties of the partial join operation. When central nodes in each of the constituent networks are joined, the diameter of the resulting network is at most max $\{d_1, d_2, 1 + r_1 + r_2\}$, where the $d_i$ and $r_i$ are the diameter and radii, respectively, of the constituent networks. However, the vertex vulnerability of the combination is unacceptable, being at most the number of central nodes in one of the constituents. Vulnerability can be reduced by adding edges. In the limit, one obtains the join in which every vertex of one constituent is adjacent to every vertex in the other. In this case, the resulting network is no more vulnerable than its weakest constituent.

Clearly, there are trade-offs between the size (measured by numbers of vertices and/or edges), and the diameter and vulnerability of the resulting network [13]. For example, if each of the two networks to be combined has $n$ vertices and $l$ edges (with diameter $d$ and vulnerability $c$), then the Cartesian product has $n^2$ vertices, $2nl$ edges, diameter $2d$, and vulnerability $cn$. This gives a network with relatively small diameter, low vulnerability, but at the expense of replicating a large number of nodes and edges.

The graph operation to be used in combining networks will depend on operational requirements as well as the desired internal structure of the combined network. The security needs of a social network, for example, might dictate a large number of links between subnets to ensure redundant paths between nodes. Detailed analysis of special cases of networks in specific operational environments is needed to determine optimal growth strategies.

## Acknowledgments

Research was sponsored by the U.S. Army Research Laboratory and the U.K. Ministry of Defence and was accomplished under Agreement Number W911NF-06-3-0001. The views and conclusions contained in this document are those of the author(s) and should not be interpreted as representing the official policies, either expressed or implied, of the U.S. Army Research Laboratory, the U.S. Government, the U.K. Ministry of Defence, or the U.K. Government. The U.S. and U.K. Governments are authorized to reproduce and distribute reprints for Government purposes notwithstanding any copyright notation hereon.

## References

1. Agrawal, D.P., Chen, C. and Burke, J.R. Hybrid graph-based networks for multiprocessing. *Telecom. Syst.*, 10(1), pp. 107–134, 1998.

2. Atay, F.M. and Biyikoglou, T. Graph operations and synchronization of complex networks. *Phys. Rev. E*, 72, 016217, pp. 1–7, 2005.

3. Barefoot, C.A., Entringer, R. and Swart, H., Vulnerability in graphs—A comparative survey. *J. Combin. Math. Combin. Comput.* 1, pp. 13–21, 1987.

4. Bent, G., Dantressangle, P., Vyvyan, D., Mowshowitz, A. and Mitsou, V. A dynamic distributed federated database. *Second Annual Conference of ITA*, Imperial College, London, September 2008.

5. Bermond, J.-C., Delorme, C. and Farhi, G. Large graphs with given degree and diameter. *Annal. Discrete Math.*, 13, pp. 23–32, 1982.

6. Dehmer, M. and Mowshowitz, A. A history of graph entropy measures. *Inf. Sci.*, 181, pp. 57–78, 2011.

7. Godsil, C. and McKay, B. A new graph product and its spectrum. *Bull. Austral. Math. Soc.*, 18, pp. 21–28, 1978.

8. Imrich, W. and Klavžar, S. *Product Graphs: Structure and Recognition.* New York: Wiley, 2000.

9. Jung, H.A. On a class of posets and the corresponding comparability graphs. *J. Combin. Theory B* 24, pp. 125–133, 1978.

10. Kirlangic, A. A measure of graph vulnerability: Scattering number. *Int. J. Math. Math. Sci.*, 30(1), pp 1–8, 2002.

11. Kratsch, D., Kloks, T. and Muller, H. Measuring the vulnerability for classes of intersection graphs. *Discrete Appl. Math.* 77, pp. 259–270, 1997.

12. Mamut, A. and Vumar, E. Vertex vulnerability of kronecker products of complete graphs. *Inf. Process. Lett.*, 106(6), pp. 258–262, 2008.

13. Miller, M. and Širáň, J. Moore graphs and beyond: A survey of the degree-diameter problem. *Electron J. Comb.*, Dynamic survey D14, pp. 61, 2005.

14. Mowshowitz, A. and Bent, G. Formal properties of distributed database networks. *Proceedings of First Annual Conference of ITA*, College Park, MD, September 2007.

15. Potočnik, P. and Šajna, M. On almost self-complementary graphs. *Discrete Math.*, 306, 107–123, 2006.

16. Zhang, S., Li, X. and Han, X. Computing the scattering number of graphs. *Int. J. Comput. Math.*, 79(2), pp. 179–187, 2002.

# Chapter 8

## Review of Structures and Dynamics of Economic Complex Networks: Large-Scale Payment Network of Estonia

Stephanie Rendón de la Torre and Jaan Kalda

CONTENTS

## 8.1   Introduction

The neologism *econophysics* was first coined by H. Eugene Stanley in a Statphys conference in 1995 held in Kolkata, India. Mantegna and Stanley [1] defined *econophysics* as a multidisciplinary field that denotes the activities of physicists who work on economic problems in order to test a variety of new conceptual approaches derived from physical sciences. Much has been studied and developed in this area since then and even before then, mainly originated from models of statistical mechanics. Similarly, problems related with distributions of income, wealth, and economic returns in financial markets have been already addressed in research papers, and mostly these topics are related with the insufficiency to explain non-Gaussian distributions and scaling properties empirically detected by means of traditional economic theoretic approaches. Some of the most relevant outcomes of the research accomplished in topics of econophysics are related with detection and explanation of power-law tails in the distribution of different types of financial data, the existence of certain underlying universalities in the behavior of individual market agents, and the detection of similarities between financial time series and natural phenomena.

In recent years, a part of the main focus of research has tilted towards the discovery and understanding of the underlying financial, social, and economic systems' structures through the use of the tools of complex networks science. In this context, the network approach has two sources of origin: one source originates from economics, finance, and sociology, while the second source originates from computer science, physics, complexity, and mathematics. The convergence point of both sources of origin attempts to combine economy and complex systems studies, and this approach can be translated into a graph representation of economic systems in order to study how interactions among the components of the graph occur whatsoever the nature of the relations between the components is.

Network science is an interdisciplinary active field of research that originates from the mathematics branch of graph theory, and it has been extended into different directions including towards economics, statistical mechanics, computer science, neuroscience, sociology, transportation, ecological systems, and biology. With complex networks, it is possible to describe the structure of any system, when the system is suitable to be represented as a graph.

"Complexity" may refer to the quality of a system or to a quantitative characterization of a system. As a quality of the system, it refers to what makes the system complex and it has something to do with the ability to understand a system; it refers to the existence of emergent properties, which appear as a consequence of the interactions of the components of the system [2]. An example of a property that emerges as a consequence of global organizational structure of a network is the "small-world" property, which is characterized by small average path length and a high number of triangles in the network. In the second definition of complexity, this term is used as a quantity when referring to something that is more complicated than other thing; it refers to the quantity of information needed to specify the system. For real-world networks, a huge amount of information is needed to describe a system, such as the number of nodes, links, degree correlations, degree distributions, clustering coefficients, diameter, betweenness, centralities, community structure, average or shortest paths, communication patterns, and other quantities. In the case of random networks, the only information needed to describe their structure is the number of nodes and the probabilities for linking pairs of nodes. The network representation of real networks is called "complex networks" because of two reasons. Firstly, because there are characteristics that arise as a consequence of the global topological organization of the system, and secondly because these structures cannot be trivially described like in the cases of random or regular graphs [3].

The theoretical framework behind complex networks is continuously developing, advancing at a fast pace and has already made significant progress towards unraveling the organizing principles governing complex networks structures and their dynamics. Studies related with: topological features, dynamical aspects, community detection, network phenomena, and particular properties of networks have been the focus of attention of extensive research in the last couple of decades [4–9].

Networks play an important role in a wide range of economic and social phenomena, and the use of techniques and methods from graph theory has permitted economic network theory to expand the knowledge and insights into such phenomena in which the embeddedness of individuals or agents in their social or economic interrelations cannot be neglected [10]. For example, Souma et al. [11] studied a shareholder network of Japanese companies where the authors analyzed the companies' growth through economic networks. Other examples of interesting applications of complex networks in economics are provided by the regional investment or ownership networks where European company-to-company investment stocks show power-law distributions that allow predicting the investments that will be received or made in specific regions, based on the connectivity and transactional activity of the companies [12,13]. Nakano and White [14] have shown that analytic concepts and methods related with complex networks can help to uncover structural factors that may influence the price formation for empirical market-link formations of economic agents. Reyes et al. [15] used a weighted network analysis focused on using random walk betweenness centrality to study why high-performing Asian economies have higher economic growth than Latin-American economies in the last years. Network-based approaches are very useful and provide a means by which to monitor complex economic systems and may help on providing better control in managing and governing these systems. Other interesting line of research is related with network topology as a basis for investigating money flows of customer-driven banking transactions. A few recent papers describe the actual topologies observed in different financial systems [16–19]. Some other works have focused on economic shocks and robustness in economic complex networks [20,21].

## 8.2  Summary

Networks can be studied from different points of view, for example, from a local, global, or mesoscale perspective. The contribution of this chapter is to explore these approaches using different methodologies with the goal of studying general and particular properties of networks through the analysis that consists of different experiments on a unique, interesting, and particular economic network. This is a review of our research on the structures and characteristics of the large-scale Estonian network of payments [22–24]. In this novel and unique economic network, the nodes represent Estonian companies and the links represent payments done between the companies. Mainly, we focus on the analysis of:

a. Global and local topology

b. Community detection and structure

c. Fractal and multifractal properties

Our data set was obtained from Swedbank's databases. Swedbank is one of the leading banks in the Nordic and Baltic regions of Europe. The bank operates actively in Estonia, Latvia, Lithuania, and Sweden. All the information related to the identities of the nodes is very sensitive and thus will remain confidential and unfortunately cannot be disclosed. The data set is unique in its kind and very interesting since ~80% of Estonia's bank transactions are executed through Swedbank's system of payments, therefore, this data set reproduces fairly well the transactional trends of the whole Estonian economy, and hence we use this data set as a proxy of the economy of Estonia. Such data set comprises domestic payments (company-to-company electronic transactions) of year 2014. The network consists of 16,613 nodes, 2,617,478 payment transactions, and 43,375 links.

## 8.3   Topologic Structure and Components: Analytic Metrics

In this sub-section of the chapter, we focus on analyzing some interesting structural properties of our network, with a special focus on topologic components. Graph theory definitions not introduced in this chapter can be found in [25,5].

A random network is the most basic model of all network formations, and it is based on the assumption that a fully random process is responsible for the structure of the links in a network. The properties of random network models [26] provide rich insight of the characteristics and features that many economic and social networks share. Such models are useful benchmarks to compare empirical networks in order to be able to identify the elements that are a result of randomness and the ones that can be rooted to other factors. Some properties of random networks that are useful for studying general networks are, for example, the distribution of links across nodes, connectivity in terms of paths, distances within networks, shortest-average paths, diameter, and etcetera.

A graph is a mathematical and symbolic representation of a network and of its connectivity. A simple undirected graph $G$ is a set of vertices $V$ connected with edges $E$, therefore, $G = (V, E)$. A graph is defined by the structural information contained in its adjacency matrix. A network may have an arbitrary large amount of additional information on top of it: for example, edges can have attributes such as capacity or weight, or it may be a function of other variables. Also, in a network, the vertices are called nodes and the edges are called links. Network terminology is generally used when the links transport or send something between the nodes (like in social, computer, biological, transport, or economic networks).

There are several ways to define our network of payments and in this study we consider more than one definition. In the first definition, we mapped an undirected graph, a symmetric payment adjacency matrix $A_{N \times N}$, where $N$ is the total number of nodes in the network, then $a_{ij}^{u} = a_{ji}^{u}$ and $a_{ij}^{u} = 1$. Otherwise, $a_{ij}^{u} = 0$ if there is no transaction between companies $i$ and $j$.

The links can also represent directions, where the links follow the flow of money. The second definition is a directed graph where the links follow the flow of money, such that a link is incoming to the receiver and outgoing from the sender of the payment. For this case, we have two more matrices, one for the in-degree case and another one for the out-degree case. The choice of the definition of the matrix representation depends on the focus of the analysis.

The most basic properties of a network are the number of nodes $N$ and the overall number of links $k$. The number of nodes defines the size of the network, while the number of links relative to the number of possible links defines the connectivity of a network. Connectivity ($p$) is the unconditional probability that two nodes are connected by a direct link. For a directed network, connectivity is defined as follows:

$$(p) = \frac{k}{n(n-1)}. \tag{8.1}$$

The connectivity of our network is 0.13, meaning that the network is sparse and 87% of the potential connections are disabled. Diameter $d$ is the maximum distance between two nodes (measured by the number of links), and this distance is equal to 29; this number is substantially higher when compared to the diameter of a random network of comparable characteristics ($d \sim 19$). The difference between the diameter number of our network and a comparable random network is substantially high, and it could be explained by the preferred money paths that nodes have in our network. Preferred money paths means that some companies have specific preferences when considering the counterparties they transact with. Intuitively, this makes sense because for a company it is important to choose carefully which counterparties become trading partners, clients, service providers, or suppliers and which ones not. Usually, this decision is based upon determined factors such as geographical location, goals affinity, cost policies, future joint ventures, legal agreements, nature of the business, or any other reasons, and it is interesting to notice how this particular feature can be observed through the comparison of the connectivity of our network and a random network.

A path is a sequence of nodes such that each node is linked to the next one along the path by a link. A path consists of $n+1$ nodes and $n$ links. A path between nodes $i$ and $j$ is an ordered list of $n$ links. The length of this path is $n$. The path length of all node pairs could be represented in the form of a distance matrix. The average path length is the average of the shortest path lengths across all node pairs in the network.

Other simple quantity that can be observed in a network is the number of nodes of a given degree. The degree of a node is the number of neighbors of that node and is defined as

$$k_i = \sum_{j \in \zeta(i)} a_{ij}, \tag{8.2}$$

the sum runs over the set $\zeta(i)$ of neighbors of $i$. For example: $\zeta(i) = \{j \mid a_{ij} = 1\}$.

The average degree of a network is the number of links divided by the number of nodes and is defined as

$$\langle k \rangle = \frac{1}{n} \sum k^0 = \frac{1}{n} \sum k^d = \frac{m}{n}, \tag{8.3}$$

where $m$ is the number of links and $n$ is the number of nodes. The average degree of our network is 20.

In a directed network, there are two important characteristics of a node: the number of links that end at a node and the number of links that start from the node. These quantities are known as the out-degree $k_o$ and the in-degree $k_d$ of a node, and we define them as

$$k_d = \sum_{j \in \zeta(i)} a_{ij}^d, \quad k_o = \sum_{j \in \zeta(i)} a_{ij}^o. \tag{8.4}$$

Also, it is possible to categorize networks by the degree distributions shown in their tails. In general, real-world networks are very different compared with random networks, when referring to their degree distributions. Random networks commonly show Poisson distributions, while real-world networks might have long tails in the right part of the distribution with values that are far above the mean. Measuring the tail of the distribution of the degree data could be achieved by building a plot of the cumulative distribution function. In real-world networks, it is common to find distributions that follow power laws in their tails:

$$P(k) \sim \sum_{k'=k}^{\infty} k'^{-\gamma} \sim k^{-(\gamma-1)}, \tag{8.5}$$

where $\gamma$ is the scaling exponent of the distribution and the degree distribution $P(k)$ is the probability that the degree of a node is equal to $k$. This type of distribution is called scale-free and networks with such degree distributions are referred to as scale-free networks. Such distributions have no natural scale and the functional form of the distribution remains unchanged within a multiplicative factor under a rescaling of the random variables. Previous studies [27,28] have shown that in large-scale-free networks, independently of the system and the origin of the components, the probability $P(k)$ that a node in the network interacts with $k$ other links decays as a power law, suggesting that there is a tendency for large networks to self-organize into a scale-free state. A degree distribution with power laws is a characteristic commonly seen in complex networks such as in the World Wide Web network, protein interaction networks, phone calls networks, food webs networks, citation networks, actors-movies networks, and it also appears in systems of payments from different banks around the world [17–19].

Complex networks can be classified as homogeneous or heterogeneous depending on their degree distributions. Homogeneous networks are identified by degree distributions that follow an exponential decay. In these networks, the distribution peaks at an average $k$ and then decays exponentially for large values of $k$, such as the distributions formed in the random graph model [26] and the small-world model [9] where each node has approximately the same number of links $k$, a normal distribution and the majority of the nodes has an average number of connections and only few or none of the nodes have either some or lots of connections. In heterogeneous large networks or scale-free networks, the degree distribution decays as a power law with a characteristic scale. The degree distribution follows a Pareto form of distribution where many nodes have few links and few nodes have many links, therefore, highly connected nodes are statistically significant in scale-free networks.

Figure 8.1a shows the cumulative degree distribution of the Estonian network of payments (undirected). A straight line was added as eye guideline.

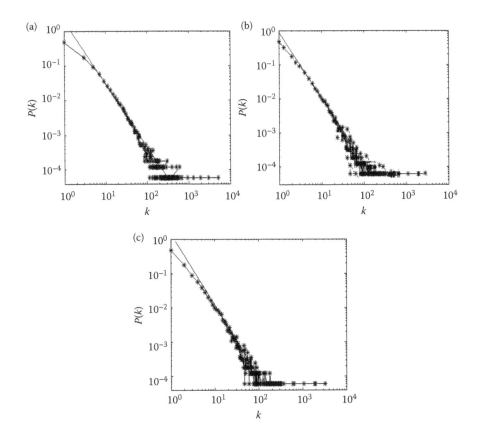

**FIGURE 8.1:** (a) Empirical degree distribution for the connectivity network of the Estonian network of payments. $X$-axis is the number of $k$ degrees and $Y$-axis is $P(k)$. (b) out-degree distribution of the network, $P(k) \sim k^{-2.39}$. (c) Empirical in-degree distribution $P(k) \sim k^{-2.49}$.

The distribution in Figure 8.1a follows a power law with the following scaling exponent:

$$P(\geq k) \propto k^{-2.4}. \tag{8.6}$$

Figure 8.1b shows the out-degree distribution and Figure 8.1c shows the in-degree distribution of our network. In all the distributions, we found regions that can be fitted by power laws, and this implies that the network has a scale-free structure.

Another interesting and fundamental metric of complex networks is the clustering coefficient of a node. It represents the probability that any two neighbors of a node are connected; it is the density around a node. In our study, it indicates whether or not there is a link between two companies that have a common third business party.

$$C(i) = \frac{1}{k_i(k_i - 1)} \sum_{j \neq k} a_{ij} a_{jk} a_{ik}. \tag{8.7}$$

The average clustering coefficient is the mean of the clustering coefficients $\langle C \rangle$ of all the nodes. In our network, the average clustering coefficient is 0.18, and this suggests there is cliquishness in the network. This means that two companies that are trading partners with a third one, have an average probability of 18% of being trading partners with another than the probability than any two other companies randomly chosen have. For visualization purposes, Figure 8.2 displays the distribution of the clustering coefficient of our network. As seen in the plot, there are a high number of unlinked neighbor nodes (45% of the nodes) that might be explained by the large number of nodes with degrees equal to 1 which appear frequently in scale-free networks.

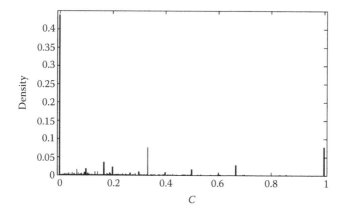

**FIGURE 8.2:** Distribution of the clustering coefficient of the Estonian network of payments.

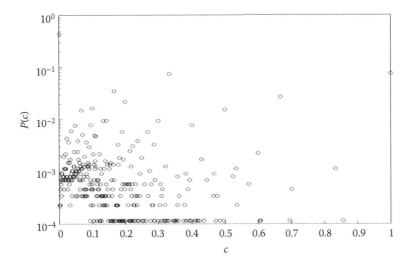

**FIGURE 8.3:** Probability distribution of the clustering coefficient of the Estonian network of payments.

We use the set of clustering coefficients of node $i$ to construct a probability distribution. Figure 8.3 shows the probability distribution of clustering coefficients of our network. As observed in the plot, the irregularity of the clustering coefficients is noticeable.

Compared to other real-world networks, such as the U.S. Federal Reserve Bank network of payments [18], the film-actor network [9], or the metabolism reactions network [28,29], the average clustering coefficient in our network is low.

As it was mentioned earlier, the most basic model of networks is the random network model $G(n, p)$ developed by Erdös and Rényi [26] and this model has two parameters: $n$ and $p$ ($n$ is the number of nodes of the graph and $p$ is the probability to link). The model works under the assumption that there could be a link $i - j$ between two nodes $i$ and $j$, and this assumption holds no matter if the nodes had a common neighbor node before the link was formed. The outcome of the model is the generation of random network graphs with a low clustering coefficient and a low variation in the degrees of the nodes. A random network cannot capture the decreasing nature of the clustering coefficient of the nodes with increase in the node degree, because the clustering coefficient of the nodes in this type of network is totally independent of the node degree and is equal to the probability of a link between any two nodes [7].

The general characteristics and statistics of the Estonian network of payments are listed in Tables 8.1 and 8.2. Regarding other statistical measures of the Estonian network of payments, as per Table 8.2, the average shortest path length $l$ is equal to 7.1 (calculated with Dijkstra's algorithm). Our network is a "small world" with 7.1° of separation, meaning that in average any company

**TABLE 8.1:**   Network's characteristics

| | |
|---|---|
| Companies analyzed | 16,613 |
| Total number of payments analyzed | 2,617,478 |
| Value of transactions | 3,803,462,026[a] |
| Average value of transaction per customer | 87,600[a] |
| Max value of a transaction | 121,533[a] |
| Min value of a transaction (aggregated in whole year) | 1000[a] |
| Average volume of transaction per company | 60 |
| Max volume of transaction per company | 24,859 |
| Min volume of transaction per company (aggregated in whole year) | 20 |

[a]All money quantities are expressed in monetary units and not in real currencies in order to protect the confidentiality of the data set. The purpose of showing monetary units is to provide a notion of the proportions of quantities and not to show exact amounts of money.

**TABLE 8.2:**   Summary of statistics

| Statistic | Value | Components | # Nodes |
|---|---|---|---|
| $N$ | 16,613 | GCC | 15,434 |
| Number of payments | 2,617,478 | DC | 1179 |
| Undirected Links | 43,375 | GSCC | 3987 |
| $\langle k \rangle$ | 20 | GOUT | 6054 |
| $\gamma^o$ | 2.39 | GIN | 6172 |
| $\gamma^i$ | 2.49 | Tendrils | 400 |
| $\gamma$ | 2.45 | Cutpoints | 1401 |
| $\langle C \rangle$ | 0.183 | Bi-component | 4404 |
| $\langle l \rangle$ | 7.1 | $k$-core | 1081 |
| $T$ | 0.13 | | |
| $D$ | 29 | | |
| $\langle \sigma \rangle$ (nodes) | 110 | | |
| $\langle \sigma \rangle$ (links) | 40 | | |

*Abbreviations:* $N$ = number of nodes. $\langle k \rangle$ = average degree. $\gamma^o$ = scaling exponent of the out-degree empirical distribution. $\gamma^i$ = scaling exponent of the in-degree empirical distribution. $\gamma$ = scaling exponent of the connectivity degree distribution. $\langle C \rangle$ = average clustering coefficient. $\langle l \rangle$ = average shortest path length. $T$ = connectivity %. $D$ = Diameter. $\langle \sigma \rangle$ = average betweenness. GCC = Giant Connected Component. DC = Disconnected Component. GSCC = Giant Strongly Connected Component. GOUT = Giant Out-Component. GIN = Giant In-Component.

can be reached by other company in just a few links. Also, our network showed low connectivity ($C = 0.13$) but at the same time the network is densely connected. This characteristic is in line with the fact that there are companies that act as hubs and lead to short distances between the other companies.

## 8.3.1   Robustness of the network

In complex networks, some nodes are essential while others are not, and identifying them is a critical task for many situations. The most essential nodes are those which if removed from the network, would cause the whole system to collapse. In order to have a deeper understanding on how the network is likely to behave as a whole in the presence of perturbations, we will address the next question: if a portion of nodes were removed, would the structure of the network become divided into disconnected clusters? How will the network respond to an actual removal of nodes? There are many approaches on how to tackle this problem and locate the "key nodes" in the network, or on how to calculate the optimal percolation threshold of nodes that would break the network into disconnected clusters. Morone and Makse [30] designed an approach that has proven to perform better than other heuristic methods (such as high degree node, $k$-core, closeness, and eigenvector centralities). Morone and Makse's algorithm optimizes a measure that can reflect the collective influence effect that arises when taking into account the entire influential set of nodes at once. This algorithm predicts a smaller set of optimal influencer nodes (the group of nodes that destroy the network if they are removed).

The collective influence of a node $CI$ is defined as the product of the node's reduced degree (the number of its nearest connections $k_i - 1$), and the total reduced degree of all nodes $k_j$ at a distance $\ell$ from it, and is represented as follows:

$$CI_\ell(i) = (k_i - 1) \sum_{j \in \partial \text{Ball}(i,\ell)} (k_j - 1), \qquad (8.8)$$

where $\ell$ is defined as the shortest path. Ball$(i, \ell)$ is the set of nodes inside a ball of radius $\ell$ around node $i$. $\partial$Ball$(i, \ell)$ is the frontier of the ball and comprises the nodes $j$ that are at a distance $\ell$ from $i$. By computing $CI$ for each node, it is possible to locate the nodes with the highest collective influence. The collective influence algorithm addresses the problem of optimal influence on the computation of the minimum structural total number of nodes that reduces the largest eigenvalue of the nonbacktracking matrix of the network.

We performed a simulation using the $CI$, where we calculate the collective influence of a group of nodes as the fall in the size of the Giant Connected Component (GCC) which would occur if the nodes of the GCC were eliminated. The GCC contains 15,434 nodes, and this quantity represents 92.8% of the nodes of the whole network. These results are displayed in Figure 8.4. The plot shows the GCC when a fraction of its nodes has been removed. The optimal percolation threshold occurs when 6.0% of the nodes are removed and that is the point where GCC$(Pc) = 0$. This result implies that there are many companies that execute a large number of payments which, in fact, have a weak influence in the economic network as a whole. The most influential companies in the network are not necessarily the most connected ones, neither those having more economic activity. A weak but smart node attack where only 6.0% of the nodes are removed destroys the whole network of payments, meaning that a few nodes maintain unified the whole network.

Scale-free networks are resilient to random removal of nodes, but are vulnerable to smart attacks. Our network is a scale-free network (with power laws in

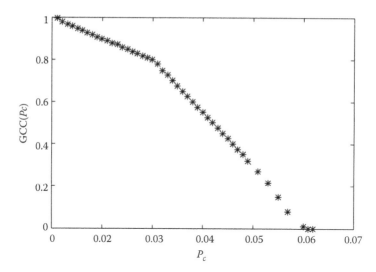

**FIGURE 8.4:** GCC of the network of payments as a function of the percolation threshold $P_c$.

the degree distribution) and its own scale-free nature makes hardly possible to destroy the network by a random removal of nodes, but if the exact portion of particularly selected nodes are removed, then the network collapses completely. This "collapse" effect has been already observed in financial systems when severe economic crisis occur and specific companies or banks declare themselves in bankruptcy and the whole system breaks down. An example is the global financial crisis of 2008 that started with the collapse of the famous investment bank Lehman Brothers, followed by Bear Sterns, UBS, and other financial entities that dragged the whole global financial system into severe liquidity problems.

## 8.4    Community Detection

Community detection analysis is essential for understanding the structure and functionality of large networks, and it also helps to expand the knowledge of the local organization of their components. Networks have sub-sections in which the nodes are more densely connected to each other than to the rest of the nodes in the network, and such sub-sections are called communities. Community detection is a graph partitioning process that provides valuable insight of the organizational principles of networks and is essential for exploring and predicting connections that are not yet observed. Thus far, recent advances of the underlying mechanisms that rule dynamics of communities in networks are limited, and this is why the achievement of an extensive and wider understanding on communities is important. Locating the underlying community structure in a network allows studying the network more easily and could provide insights into the function of the system represented by the network, as communities often correspond to functional units of systems. The study of communities and their properties also helps on revealing relevant groups of nodes, creating meaningful classifications, discovering similarities, or revealing unknown linkages between nodes. Communities have a strong impact in the behavior of a network as a whole and studying them is fundamental in order to expand the knowledge of the community structure beyond the local organization of the components of networks.

In this sub-section of the chapter, we study the overlapping community structure of our network by examining its characteristics and scale-free properties through the Clique Percolation Method (CPM) [31,32]. First, we detect communities and then we analyze the global structure of the whole network through the distribution functions of four basic quantities. In this analysis, our data set included ∼3.4 million payments from the period of October 2013 to December 2014.

The majority of previous studies on communities have essentially been devoted to the description of structures inside the communities and their applications: communities representing real social groupings [33–35] communities in a co-authorship network representing related publications of specific topics [36], protein–protein interaction networks [37], communities in a metabolic network representing cycles and functional units in biology [38], and communities in the World Wide Web representing web pages with related contents [39]. Regarding community studies on economic networks and their applications, Vitali and Battiston [40] studied the community structure of a global corporate network and found that geography is the major driver of organization within that network. Fenn et al. [41] studied the evolution of communities of a foreign exchange market network in which each node represents an exchange rate and each link represents a time-dependent correlation between the rates. By using community detection, they were able to uncover major trading changes that occurred in the market during the credit crisis of 2008. Other related economic studies have focused on the overlapping feature of communities, such as in [42,43].

Most of the algorithms for community detection can be classified as divisive, agglomerative, or optimization-based methods, and each method has specific strengths and weaknesses. Previous studies on communities based on divisive and agglomerative methods consider that structures of communities can be expressed in terms of separated groups of clusters [44], but most of the real networks are characterized by well-defined statistics of overlapping communities. An important limitation of the popular node partitioning methods is that a node must be in one single community, whereas it is often more appropriate to attribute a node to several different communities, particularly in real-world networks. An example where community overlapping is commonly observed is in social networks where individuals typically belong to many communities such as: work teams, religious groups, friendship groups, hobby clubs, family, or other similar social communities. Moreover, members of social communities have their own communities and this, in turn, results in a very complex web of communities [32]. The phenomenon of community overlapping has been already noticed by sociologists but has barely been studied systematically for large-scale networks [31,45].

Networks have sections in which the nodes are more densely connected to each other than to the rest of the nodes in the network, and such sub-sections are called communities. Communities might exist in networked systems of different nature, such as economics, sociology, biology, engineering, politics, and computer science. There is no unique definition of community in the existing literature. Definitions change depending on the author and the type of study, and precisely one of the core issues in community detection is the lack of a unified definition of what is a community. We use the CPM definition because such algorithm allows overlapping nodes among communities, a condition that arises when a node is a member of more than one community. In economic systems, the nodes could frequently belong to multiple communities; therefore, forcing each node to belong to a single community could result into a misleading characterization of the underlying community structure.

An overlapping community graph is a network that represents links between communities. In our study, the nodes represent communities and the links represent shared nodes between communities. CPM is based on the density of links and the definition of community for this algorithm is local and it is not too restrictive. Overlapping communities arise when a node is a member of more than one community. CPM is based on the assumption that a community comprises overlapping sets of fully connected sub-graphs and detects communities by searching for adjacent cliques. A clique is a complete (fully connected) sub-graph. A $k$-clique is a complete subgraph of size $k$ (the number of nodes in the sub-graph). Two nodes are connected if the $k$-cliques that represent them share $k-1$ members. The method begins by identifying all cliques of size $k$ in a network. When all the cliques are identified, then a $N_c \times N_c$ clique–clique overlapping symmetric matrix $O$ can be built, where $N_c$ is the number of cliques and $O_{ij}$ is the number of nodes shared by cliques $i$ and $j$ [46]. This overlapping matrix $O$ encodes all the important information needed to extract the $k$-clique communities for any value of $k$. In the overlapping matrix $O$, rows and columns represent cliques and the elements are the number of shared nodes between the corresponding two cliques. Diagonal elements represent the size of the clique and when two cliques intersect they form a community. For certain $k$-values, the $k$-clique communities form such connected clique components in which the nearby cliques are linked to each other by at least $k-1$ adjacent nodes. In order to find these components in the overlapping matrix $O$, one should keep the entries of the overlapping matrix which are larger than or equal to $k-1$, set the others to zero and finally locate the connected components of the overlapping matrix $O$. The formed communities are the identified separated components.

For our method, it is important to select a proper parameter $k$. This parameter affects the constituents of the overlapping regions between communities. The larger the parameter $k$ is, the less the number of nodes which can arise in the overlapping regions. When $k \to \infty$, the maximal clique network is identical to the original network and no overlap is identified. The choice of $k$ depends on the network. It is observed from many real-world networks that the typical value of $k$ is often between 3 and 6 [47].

Figure 8.5 shows a plot of the number of communities and the average size of the communities at different $k$-values. As $k$ increases the number of communities decreases while the size of communities increases fast. When $k$ decreases the number of communities increases fast and the size of the communities remains low. In order to obtain the optimal value $k$ for our network, we tested different values ranging from 3 to 10 and *a posteriori* we found that the optimal number is $k = 5$. When $k < 5$ a high number of communities arises and the partitions become very low; when $k > 5$ a lower number of communities arises and the partitions become unreal. At the level of $k = 5$, we obtain the richest partition with the most widely distributed cluster sizes set for which no giant community appears.

For visualization purposes and in order to draw a readable map of the network, Figure 8.6 shows a graphic view of a representative section of the overlapping network of communities where big and small communities can easily be distinguished. This image depicts 25 overlapping communities and each colored circle represents a node which, in turn, represents an overlapping community. The links represent the shared nodes between the communities. The size of the nodes characterizes the size of each community. For example, the big node in the middle represents a community with 61 companies.

The usefulness of identifying the communities within this network lies in how this information could be used in a practical scenario. The output of the

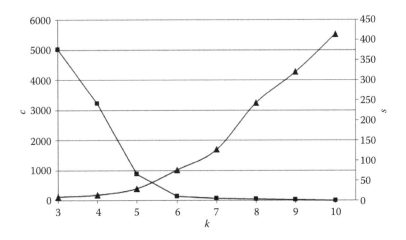

**FIGURE 8.5:** Plot of average community sizes "$s$" and number of communities "$c$" as $k$ increases. Squares represent the number of communities and triangles represent the size of the communities.

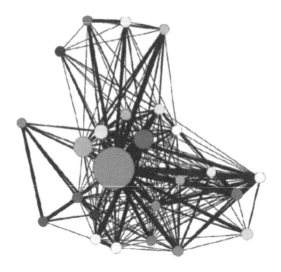

**FIGURE 8.6:** Visual representation of a section of the overlapping network of communities (Estonian network of payments). Nodes represent communities and the links represent shared nodes between communities.

community analysis could be used for targeted marketing. For example, it could be useful when integrating criteria for creating target groups of companies or customers to whom certain products or lines of products would be offered. Companies included in the same community would be located in the same target group and later on after a product offer is made it would be possible and interesting to assess the contagion effect of the product acquisition among companies of the same communities who received the offer. Another useful application is that the output of the analysis could help on creating customer-level segmentations or marketing profiles. Knowing the community (or communities) where a company or customer belongs to could be one of the drivers for creating a customer profile or grouping level. An alternative usage of the results of the community detection analysis is in predictive analytics for building churn models. Churn models usually define a measure of the potential risk of a customer cancelling a product or service and provide awareness and metrics to execute retention efforts against churning. Additionally, community detection analysis could be used as input for product affinity analysis and recommender systems. Affinity analysis is a data mining technique that helps to group customers based on historical data of purchased products and is used for cross-selling product recommendations. Another useful and immediate application is in product acquisition propensity models. These models calculate customers' likelihood to acquire a product based on a myriad of variables and the output of the overlapping community analysis could be input for such propensity models and support efficiency in sales processes.

### 8.4.1   Structure of communities

We studied the global community structure of our network by inspecting the distribution functions of four elemental quantities: community size $P(s)$, overlap size $P(s_o)$ community degree $P(d)$, and membership number $P(m)$. The distributions of such quantities are shown in Figure 8.7a–d which show important statistics that describe the community structure of our network. In general, nodes in a network can be characterized by a membership number which represents the number of communities a node belongs to. This means that, for example, any two communities may share some of their nodes which correspond to the overlap size between those communities. There is also a network of communities where the overlaps are the links and the communities are the nodes, and the number of such links is called community degree. The size of any of those communities is defined by the number of its nodes.

Figure 8.7a displays the cumulative distribution function of the community size $P(s)$. This is the probability for a community of having a community size higher or equal to $s$, calculated over different points in time $t$ (where $t$ is the time expressed in months, $t = 1$ is October, $t = 2$ is November, etc.). The overall distribution of community sizes resembles a power law $P(s) \propto s^a$, where $a$ is the scaling exponent, and a power law is valid nearly over all times $t$, suggesting

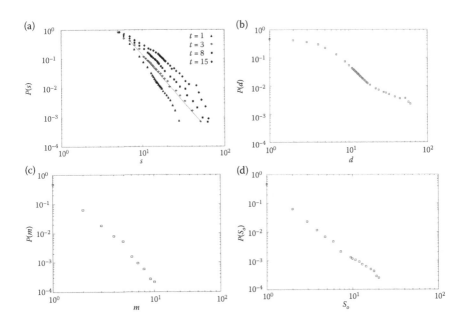

**FIGURE 8.7:** (a) Cumulative community size distribution at different times $t$. (b) Cumulative distribution of community degrees $d$. (c) Cumulative distribution function of the membership number $m_i$. (d) Cumulative distribution function of the overlap size $s_o$.

there is no characteristic community size in the network. The sizes of the communities on $t = 1$ are smaller than in the rest of the months; as time increases the size increases, particularly the size of the largest communities. The distribution at different moments in time follows similar decaying patterns, but in general, the scaling tail is higher as $t$ increases. The shapes of the power laws observed in the community size distributions of Figure 8.7a suggest there is no characteristic community size in the network. A fat tail distribution implies that there are numerous small communities coexisting with few large communities [48,49]. The scaling exponent when $t = 3$ is $-2.8$ (included for eye guideline) and Equation 8.9 is

$$P(s) \propto s^{-2.8}. \tag{8.9}$$

In a network of overlapping communities, the overlaps are represented by the links and the number of those links is represented by the community degree $d$. Then, the degree $d$ is the number of communities another community overlaps with. Figure 8.7b shows the cumulative distribution of the community degrees in the network. There are some outstanding community degrees by the end of the tail and these include communities that cluster the majority of the biggest customers in the network. The central part of the distribution decays faster than the rest of the distribution. There is an observable curvature in the log–log plot, however no approximation method fitted the distribution. This plot shows that the maximum number of degrees $d$ is 63 and corresponds to a relatively small quantity of nodes.

The membership number $m_i$ represents the number of communities a node $i$ belongs to while the overlap size $s_o$ is the number of nodes that two communities share. Figure 8.7c and d show the distributions of both measures, respectively. Both distributions seem to have a power-law behavior and indicate that there is no characteristic scale in the overlapping size or in the membership sizes. Regarding the overlap size, the range to which the communities overlap with each other is also an important property of our network.

As shown in Figure 8.7c, the largest membership number found in the network was 10, meaning that a company can belong to maximum 10 different communities simultaneously. This plot shows that the fraction of nodes that belong to many different communities is quite small, while the fraction of nodes belonging to at least 1 community is high. For example, when $m = 1$ the percentage of nodes that belong to at least one community is 50%, while the percentage of nodes that belong simultaneously to 10 communities ($m = 10$) is extremely small. However the rest of the communities belong to at least 2 or more communities.

In our previous study [22], we found scale-free properties in the degree distributions of the Estonian network of payments and it is very interesting to observe that the scale-free property is also preserved at a higher level of organization where overlapping communities are present.

## 8.5 Multifractal Networks

In the late 1960s Benoit Mandelbrot was the first to coin the term "fractal" and he also was the first one in describing the fractal geometry of nature [27], and since then the fractal approach has been widely spread and used in extensive research studies related with the underlying scaling of different complex structures, including networks.

Whether if a single fractal scaling spans or not all the constituents or areas of a system, is a fundamental issue that helps on distinguishing when a system is multifractal or just fractal. One scaling exponent is enough to characterize completely a monofractal process. Monofractals are considered as homogeneous objects because they have the same scaling properties branded by one singularity exponent. Instead, a multifractal object requires several exponents to characterize its scaling properties. Multifractals are inherently more complex and inhomogeneous than monofractals and portray systems with high variations or fluctuations that originate from specific characteristics.

Fractal and multifractal analysis can help to reveal the structure of all kinds of systems in order to have a better understanding of them. In particular, both approaches have many different interesting applications in economy. An interesting line of research is related with the relevance and applicability of fractal and multifractal analysis in social and economic topics. Inaoka et al. [17] showed that the study of the structure of a banking network provides useful insight from practical points of view. By knowing and understanding the structure and characteristics of banking networks (in terms of transactions and their patterns), a systemic contagion could potentially be prevented. In their study, these authors showed that the network of financial transactions of Japanese financial institutions has a fractal structure. Regarding social studies, Lu et al. [50] showed the importance of road patterns for urban transportation capacity based on fractal analysis of such network. In this study, the authors were able to link the fractal measurement with city mass measurements. A few recent studies have focused on the analysis of the changes of multifractal spectra across time to assess changes in economy during crisis periods [51]. Some other studies have focused on gathering empirical evidence of the common multifractal signature in economic, biological, and physical systems [52].

Fractal analysis helps to distinguish global features of complex networks, such as the fractal dimension. However, the fractal formalism is insufficient to characterize the complexity of many real networks which cannot be described by a single fractal dimension. Furuya and Yakubo [6] demonstrated analytically and numerically that fractal scale-free networks may have multifractal structures in which the fractal dimension is not sufficient to describe the multiple fractal patterns of such networks, therefore, multifractal analysis rises as a natural step after fractal analysis.

Multifractal structures are abundant in social systems and in a variety of physical phenomena. Inhomogeneous systems which do not follow a self-similar scaling law with a sole exponent could be multifractal if they are characterized by many interweaved fractal sets with a spectrum of various fractal dimensions. Multifractal analysis is a systematic approach and a generalization of fractal analysis that is useful when describing spatial heterogeneity of fractal patterns [53]. Multifractal network analysis requires taking into account a physical measure, like the number of nodes within a box of specific size in order to analyze how the distribution of such number of nodes scales in a network as the size of the box grows or reduces. In the last years, numerous algorithms for calculating the fractal dimension and studying self-similar properties of complex networks have been developed and tested extensively [31,54–57]. Song et al. [58] developed a method for calculating the fractal dimension of a complex network by using a box-covering algorithm and identified self-similarity as a property of complex networks [59]. Additionally, several algorithms and studies on multifractal analysis of networks have been proposed and developed recently [60–63].

In this sub-section of the chapter, we analyze fractal and multifractal properties of the large-scale economic network of payments of Estonia. We perform a fractal scaling analysis by estimating the fractal dimension of our network and its skeleton. Then, we study the multifractal behavior of the network by using a sandbox algorithm for complex networks to calculate the spectrum of the generalized fractal dimensions $D(q)$ and mass exponents $\tau(q)$.

### 8.5.1   Fractal network analysis

According to Song et al. [59], the box-counting algorithm is an appropriate method to study global properties of complex networks. The fundamental relation of fractal scaling is based on the box-covering method which counts the total number of boxes that are needed to cover a network with boxes of certain size. The box-covering method is equivalent to the box-counting method widely used in fractal geometry and is a basic tool for measuring the fractal dimension of fractal objects embedded in Euclidean space [64]. However, the Euclidean metric is not well defined for networks, thus we use the networks' adaptation [63] of the random sequential box-covering algorithm [65] in order to calculate the fractal dimension of our network and its skeleton. This method involves a random process for selecting the position of the center of each box. We let $N_B(r_B)$ be the minimum number of boxes needed to tile the whole network, where the lateral size of the boxes is the measure of radius $r_B$ as follows:

$$N_B(r_B) \sim r_B^{-d_B}, \tag{8.10}$$

where $d_B$ is the fractal dimension. If we measure the number of $N_B$ for different box sizes, then it is possible to obtain the fractal dimension $d_B$ by obtaining the power-law fitting of the distribution. The algorithm selects a random node at each step, and this node is the seed that will be the center of a box. Then we search the network by distance $r_B$ from the seed node and cover all the nodes that are located within that distance, but only if they have not been covered yet. Later we assign the newly covered nodes to the new box; if there are no more newly covered nodes then the box is removed. This process is repeated until all the nodes of the network belong to boxes. Before using the algorithm we calculate the skeleton of our network.

One of the main challenges of complex network studies is the identification of critical structural features that are underneath the network's complexity. This is related with the basic concept of: the distinctive character of a whole is inside just a few of its parts, for example, in specific colors and shapes of a painting, particular notes or tunes in a song or certain keywords in a text or speech. This basic concept is also true for complex networks, where only a few parts of the whole network reflect the most important properties of it. For example, in large-scale networks, only a small number of links are critical for the network to exist as a whole. A skeleton network is generally smaller than the original and it reproduces all the fundamental properties of the whole because it contains the essence of the network. Grady et al. [66] analyzed the network of international flight connections and discovered that the skeleton network consists of just 6.76% of the original network. The skeleton network concept can be used to detect epidemic propagations of disease when indicating which individuals are key participants in a social network or it can be useful when describing ecosystems to identify the species that should not be damaged at all to avoid jeopardizing the whole network.

The concept of skeleton was first introduced by Kim et al. [67]. The skeleton is a particular type of spanning tree based on the link betweenness centrality (a simplified quantity to measure the traffic of networks) that is entrenched beneath the original network. The skeleton provides a shell for the fractality of the network and is formed by links with the highest betweenness centralities. Only the links that do not form loops are included. The remaining links from the original network which are not included in the skeleton are local shortcuts that contribute to loop formation, meaning that the distance between any two nodes in the original network may increase in the skeleton. A fractal network has a fractal skeleton beneath which is distressed by these local shortcuts but it preserves fractality. For a scale-free network the skeleton also follows a power-law degree distribution where the degree exponent might differ slightly from that of the original network. When studying the origin of fractality in networks, actually the skeleton is more useful than the original network itself due to its unsophisticated and simplistic tree structure [68]. In general, the skeleton preferentially collects the sections of the network where betweenness is high and this preserves the structure and simplifies its complexity. Therefore, by looking at the properties of the skeleton it is easier to appreciate the topological organization of the original network.

In order to calculate the skeleton of a complex network, the link betweenness of all the links in the network has to be calculated. The betweenness centrality of

**FIGURE 8.8:** Graph representation of the skeleton of the Estonian network of payments.

a network (for a link or a node), is defined as follows:

$$b_i = \sum_{j,k \in N, j \neq k} \frac{n_{jk}(i)}{n_{jk}}, \qquad (8.11)$$

where $N$ is the total number of nodes, $n_{jk}$ is the total number of shortest paths between nodes $j$ and $k$, $n_{jk}(i)$ is the total number of shortest-paths linking nodes $j$ and $k$ that passes through the node $i$. In order to perform the fractal scaling analysis, we used Dijkstra's algorithm [69]; then we used the box-covering algorithm to calculate the fractal dimension of the network and the skeleton to compare both values.

We present a fractal scaling analysis by using the box-counting algorithm expressed in Equation 8.10 and we calculated the fractal dimension of our network and its skeleton. Figure 8.8 shows a visualization of the graph representation of the skeleton of our network. The box-covering method yields a fractal dimension $d_{Bs} = 2.32 \pm 0.07$ for the skeleton network and for the original network the fractal dimension is $d_{Bo} = 2.39 \pm 0.05$.

The comparison of the fractal scaling in our network and its skeleton structure revealed its own patterns according to the fractality of the network. Figure 8.9 shows a fractal scaling representation of our network and its skeleton, where the fractal dimension is the absolute value of the slope of the linear fit.

As seen in the plot of Figure 8.9, the respective number of boxes needed to cover both networks is very similar but not identical, actually more boxes were needed for covering the skeleton. The largest distance between any two

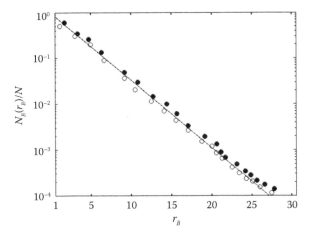

**FIGURE 8.9:** Fractal scaling representation of our network. The original network (O) and the skeleton network (●). The straight line is included for guidance and has a slope of 2.3. The analysis includes only the giant connected cluster of the network.

nodes in the network of payments is 29, while the largest distance between any two nodes in the skeleton network is 34.

### 8.5.2 Multifractal network analysis

Scale-free networks are commonly observed in a wide array of different contexts of nature and society. In the first sub-section of this chapter, we have shown that the Estonian network of payments has scale-free properties characterized by power-law degree distributions.

In general, multifractality is expected to appear in scale-free networks due to the fluctuations that occur in the density of local nodes. Tél et al. [70] introduced a sandbox algorithm based on the fixed-size box-counting algorithm [71] which was used and adapted for multifractal analysis of complex networks by Liu et al. [61]. In order to determine the multifractal dimensions of our complex network, we chose this adapted sandbox algorithm because it is precise, efficient, and practical. Moreover, a study by Song et al. [53] has shown that this algorithm gives better results when it is used in unweighted networks, and this is our case.

The fixed-size box-counting algorithm is one of the most known and efficient algorithms for multifractal analysis. For a given probability measure $0 \leq \mu \leq 1$ in a metric space $\Omega$ with a support set $E$, we consider the following partition sum:

$$Z_\varepsilon(q) = \sum_{\mu(B) \neq 0} [\mu(B)]^q, \qquad (8.12)$$

where the parameter $q \in R$, and describes the moment of the measure. The sum runs over all different non-overlapping (or non-empty) boxes $B$ of a given size $\varepsilon$ that covers the support set $E$. From this definition, it is easy to obtain $Z_\varepsilon(q) \geq 0$ and $Z_\varepsilon(0) = 1$. The function of the mass exponents $\tau(q)$ of the measure $\mu$ is defined by

$$\tau(q) = \lim_{\varepsilon \to 0} \left( \frac{\ln Z_\varepsilon(q)}{\ln \varepsilon} \right). \tag{8.13}$$

Then, the generalized fractal dimensions $D(q)$ of the measure $\mu$ are defined as follows:

$$D(q) = \frac{\tau(q)}{q-1}, \quad q \neq 1, \tag{8.14}$$

and

$$D(1) = \lim_{\varepsilon \to 0} \frac{Z_{(1,\varepsilon)}}{\ln \varepsilon}, \quad q = 1, \tag{8.15}$$

where

$$Z_{1,\varepsilon} = \sum_{\mu(B) \neq 0} \mu(B) \ln \mu(B). \tag{8.16}$$

The generalized fractal dimensions $D(q)$ can be estimated with linear regression of $[\ln Z_\varepsilon(q)]/[q-1]$ against $\ln \varepsilon$ for $q \neq 1$, and similarly a linear regression of $Z_{1,\varepsilon}$ against $\ln \varepsilon$ for $q = 1$. $D(0)$ is the fractal dimension or the box-counting dimension of the support set $E$ of the measure $\mu$, $D(1)$ is the information dimension and $D(2)$ is the correlation dimension.

For a complex network, a box of size $B$ can be defined in terms of the distance $l_B$, which corresponds to the number of links in the shortest path between two nodes. This means that every node is less than $l_B$ links away from another node in the same box. The measure $\mu$ of each box is defined as the ratio of the number of nodes that are covered by the box and the total number of nodes in the whole network.

Multifractality of a complex network can be determined by the shape of $\tau(q)$ or $D(q)$ curves. If $\tau(q)$ is a straight line or $D(q)$ is a constant, then the network is monofractal; similarly if $D(q)$ or $\tau(q)$ have convex shapes, then the network is multifractal. A multifractal structure can be identified by the following signs [72]: multiple slopes of $\tau(q)$ versus $q$, non-constant $D(q)$ versus $(q)$ values and $f$ $(a)$ versus $a$ value covers a broad range (not accumulated at nearby non-integer values of $a$).

Firstly, we calculate the shortest-path distance between any two nodes in the network and map the shortest-path adjacency matrix $B_{N \times N}$ using the payments adjacency matrix $A_{N \times N}$. Then we use the shortest-path adjacency matrix $B_{N \times N}$ as input for multifractal analysis. The central idea of the sandbox algorithm is simply to select a node of the network in a random fashion as the center of a sandbox and then count the number of nodes that are inside the sandbox. Initially, none of the nodes has been chosen as a center of a box or as a seed. We set the radius $r$ of the sandbox which will be used to cover the nodes in the range $r \in [1, D]$, where $D$(diameter) is the longest distance between nodes in the network and radii $r$ are integer numbers. We ensure that the nodes are chosen randomly as center nodes by reordering the nodes randomly in the whole network. Depending on the size $N$ of the network, we choose $T$ nodes in random order as centers of $T$ sandboxes; then we find all the neighboring nodes within radius $r$ from the center of each box. We count the number of nodes contained in each sandbox of radius $r$, and denote that quantity by $S(r)$. We calculate the statistical averages $[S(r)^{q-1}]$ of $[S(r)^{q-1}]$ over all the sandboxes $T$ of radius $r$. The previous steps are repeated for each of the different values of radius $r$ in order to obtain the statistical average $[S(r)^{q-1}]$ and use it for calculating linear regression.

The generalized fractal dimensions $D(q)$ of the measure $\mu$ are defined by

$$D(q) = \lim_{r \to 0} \frac{\ln [S(r)/S(0)]^{q-1}}{\ln (r/d)} \frac{1}{q-1}, \quad q \in R \tag{8.17}$$

or rewritten as

$$\ln ([S(r)]^{q-1}) \propto D(q)(q-1) \ln\left(\frac{r}{d}\right) + (q-1)\ln(S_0), \tag{8.18}$$

where $S(0)$ is the size of the network and the brackets mean taking statistical average over the random selection of the sandbox centers. We run the linear regression of $\ln([S(r)]^{q-1})$ against $(q-1)\ln(r/d)$ to obtain the generalized fractal dimensions and similarly, calculate the linear regression of $\ln([S(r)]^{q-1})$ against $\ln(r/d)$ to obtain the mass exponents $\tau(q)$. From the shapes of the generalized fractal dimension curves, we can conclude if multifractality exists or not in our network.

Linear regression is an important step to obtain the correct range of radius $r \in [r_{min}, r_{max}]$ that is needed to calculate the generalized fractal dimensions (defined by Equations 8.17 and 8.18) and the mass exponents (defined by Equation 8.13). We found an appropriate range of radii $r$ within the range of the interval located between 2 and 29 for linear regression, thus we selected this linear fit scaling range to perform multifractal analysis (we set the range of $q$-values from $-7$ to $12$).

We calculated $\tau(q)$ and the $D(q)$ curves using the sandbox algorithm by Liu et al. [61] and based upon the shapes obtained from the spectrum in Figure 8.10a and b, it can be seen that the curves are non-linear, suggesting that the network is multifractal.

In Figure 8.10b, the $D(q)$ function decreases sharply after the peak reaches its end when $q$ is $-4$. This could be interpreted as the high densities around

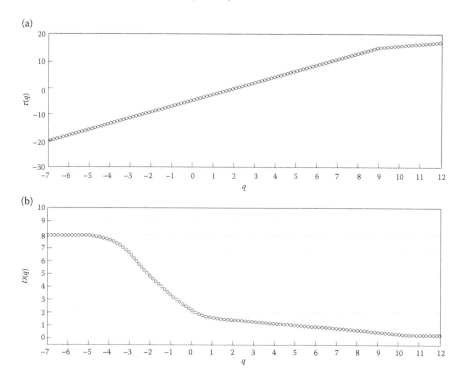

**FIGURE 8.10:** (a) Plot of mass exponents $\tau(q)$ as function of $q$. (b) Plot of generalized fractal dimensions $D(q)$ as function of $q$. Curves indicated by circles represent numerical estimations of the mass exponents and generalized fractal dimensions, respectively.

the hubs in the network. The hubs have a high number of links connected to them; therefore, the density of links around the sections near the hubs is higher than in other parts of the network. These hub nodes or important companies have a noticeable larger amount of business partners (for example: customers, suppliers, or any other business parties that interact financially) than the rest of the companies in the network have, and it is interesting to observe that this characteristic can be explored and identified by looking at the values of $D(q)$ spectra. The multifractality seen in our network reveals that the system cannot be described by a single fractal dimension suggesting that the multifractal approach provides a better characterization; hence, this means that the Estonian economy is multifractal.

The quantity $\Delta D(q)$ describes the changes on link density in our network. We use $\Delta D(q) = D(q)_{max} - \lim D(q)$ to observe how the values of $D(q)$ change along the spectrum. From Figure 8.10b, we found that $\lim D(q) = 0.37$ and $D(q)_{max} = 7.8$ and this means that $\Delta D(q) = 7.43$. A large $D(q)$ value means that the link distribution is very irregular, suggesting that there are areas of hubs where the links are densely grouped contrasting with areas where the nodes are connected with only just few links. In our network, this means that just a few companies have the role of hubs, while the rest are just small participants of the payments network. Table 8.3 shows a comparison of the maximum values of $D(q)$ in different networks.

**TABLE 8.3:** Comparison of the maximum values of $D(q)$ in different networks

| Network | Number of nodes | Highest $D(q)$ | Reference |
|---|---|---|---|
| Pure fractal network | 6222 | 2.8 | [60] |
| Small-world network | 6222 | 6.6 | [60] |
| Semi fractal network | 6222 | 3.1 | [60] |
| Sierpinski weighted fractal network | 9841 | 2.0 | [53] |
| Cantor dust weighted fractal network | 9841 | 3.2 | [53] |
| High-energy theory collaboration weighted network | 8361 | 6.0 | [53] |
| Astrophysics collaboration weighted network | 16,706 | 6.2 | [53] |
| Computational geometry collaboration weighted network | 7343 | 5.1 | [53] |
| Barabási and Albert model scale-free network | 10,000 | 3.6 | [61] |
| Newman and Watts model small-world network | 10,000 | 4.8 | [61] |
| Erdös–Rényi random graph model | 10,000 | 3.9 | [61] |
| Barabási and Albert model scale-free network | 7000 | 3.4 | [63] |
| Random network | 5620 | 3.5 | [63] |
| Random network | 449 | 2.4 | [63] |
| Protein–Protein interaction network: Human | 8934 | 4.9 | [63] |
| Protein–Protein interaction network: *Arabidopsis thaliana* | 1298 | 2.5 | [63] |
| Protein–Protein interaction network: *C. elegans* | 3343 | 4.5 | [63] |
| Protein–Protein interaction network: *E. coli* | 2516 | 4.1 | [63] |
| Small-world network | 5000 | 3.0 | [63] |
| Estonian network of payments | 16,613 | 7.8 | [22] |

## 8.6    Conclusions

Complex networks can be considered as the skeleton of complex systems and they are present in many kinds of social, economic, biological, chemical, physical, and technological systems. In this chapter, we have reviewed global properties and statistics related with the topological structure of the large-scale payments network of an entire country (Estonia) by using payments data. Additionally, we have reviewed some topics related with its community structure and moreover, we have analyzed some aspects related with multifractal and fractal properties.

In the network of Estonian payments, we found scale-free degree distributions, small-world property, low clustering coefficient, disassortative degree, and heterogeneity. Its scale-free structure indicates that a low number of companies in Estonia trade with a high number of companies, while the majority of the companies trade with only few. The clustering coefficient distribution suggests the existence of a hierarchic structure in the network. Our network is a small world with just 7° of separation. The connectivity is smaller than the overall clustering coefficient therefore, our network is not random. The diameter value suggests there is a preference among companies for particular paths of money.

We tested the robustness of the network with an approach that focuses on the collective influencer nodes. First, we located the key nodes that prevent the network of breaking into disconnected components. The simulation assumed a targeted removal of key nodes which cause a quick growth in the average shortest path length until the network was destroyed at an optimal percolation threshold of 6%, while in the random removal the damage was extremely small. This revealed the robustness of our economic network against random attacks but also revealed its vulnerability to smart attacks. The low percentage of the optimal percolation threshold reveals that the most influential companies in the network are not necessarily the most connected ones or those having more economic activity and that a small quantity of companies maintains the whole network unified.

Later, we analyzed the community structure of our network by using the CPM. We found that there are scale-free properties in the statistical distributions of the community structure, too. Size, overlap, and membership distributions follow shapes that are compatible with power laws. Power-law distributions have already appeared in this network at a global scale in the level of nodes, and in this community study we have shown that power laws are present at the level of overlapping communities, too.

An immediate application for the community detection output is that it can be used in targeted marketing activities, as input for predictive analytical models such as in product acquisition propensities, churn, product affinity analyses, for creating marketing profiles or customer segmentations and for creating customer target lists for product offering (in an effort to propagate consumer buzz effects). Further applications for community detection in similar economic networks could involve identification of patterns between companies, tracking suspicious business activities, and strengthening relationships between companies of the same community for improving performance of the whole network.

In the last part of the chapter, we presented a fractal and multifractal analysis of the network. We identified the underlying structure of the network (its skeleton) and measured the fractal dimension of the skeleton to compare it with the fractal dimension of the original network. Both fractal dimensions were similar but the fractal dimension of the skeleton was slightly smaller. We also analyzed the general multifractal structure by calculating the spectrum of the mass exponents $(q)$ and the generalized fractal dimension $D(q)$ curves, through a sandbox algorithm for multifractal analysis of complex networks. Our results indicated that multifractality exists in the Estonian network of payments, and this suggests that the Estonian economy is multifractal (from the point of view of networks). We found large values of $D(q)$ spectra, which mean that the distribution of links is quite irregular in the network, suggesting there are specific nodes which hold densely connected links while other nodes hold just a few links. This type of structure could be relevant when critical events occur in the economy that could threaten the whole network.

It is important to continue studying the structures and characteristics of economic complex networks in order to be able to understand their underlying processes and to be able to detect patterns that could be useful for predicting or forecasting events and trends. The addition of evidence through empirical studies in favor of fractality, multifractality, communities' detection, and structural properties of economic networks represents a step forward towards the knowledge on the unraveling of the complexity of economic systems.

### 8.6.1 Further applications

Regarding community structure in economic networks, a question that remains open for future research is to investigate if the similarities in communities' features amongst different complex networks arise randomly or if there are any unknown properties shared by all of them. Another interesting open line of research is to study the plausibility of predicting changes in a payment network through community detection analysis. Further applications in economic networks could involve strengthening relationships between companies of the same community to improve the performance of the whole network, targeted marketing, identification of patterns between companies, and tracking of suspicious business activities.

Further applications of multifractal studies in economic networks might involve examining the potential factors that drive the strength of the multifractal spectrum. Some applications could involve studying the origin of such factors. Another interesting line of research would be to study the patterns and the changes of the multifractal spectrum across different periods of time.

Particularly, it would be interesting to analyze such patterns during financial crisis periods for risk pattern recognition purposes. Also, it would be interesting to take into account different probability measures for such kind of multifractal analysis. Other direction of the studies could focus on building network models that attempt to forecast country money flows or potential industry growth trends based on data of transactions.

# References

1. Mantegna R.N., & Stanley H.E. 2000. *An Introduction to Econophysics: Correlations and Complexity in Finance.* Cambridge, UK: Cambridge University Press.

2. Standish R.K. 2008. *Intelligent Complex Adaptive Systems.* New South Wales, Australia: IGI Global, 105–124.

3. Estrada E. 2011. *The Structure of Complex Networks.* New York, NY: Oxford University Press.

4. Reka A., & Barabási A.L. 2002. Statistical mechanics of complex networks. *Reviews of Modern Physics*, 74(1), 47–97.

5. Dorogovtsev S.N., & Mendes J.F.F. 2003. *Evolution of Networks.* New York, NY: Oxford University Press.

6. Furuya S., & Yakubo K. 2011. Multifractality of complex networks. *Physical Review E*, 84(3), 036118.

7. Newman M.E.J. 2010. *Networks: An Introduction.* New York, NY: Oxford University Press.

8. Palla G., Barabási A.L., & Vicsek T. 2007. Quantifying social group evolution. *Nature*, 446, 664–667.

9. Watts D.J., & Strogatz S.H. 1998. Collective dynamics of small-world networks. *Nature*, 393(6563), 440–442.

10. König M.D., & Battiston S. 2009. From graph theory to models of economic networks, a tutorial in: Naimzada A.K., Stefani S., & Torriero A. (eds). *Networks, Topology and Dynamics.* Lecture Notes in Economics and Mathematical Systems. Germany: Springer, 613, 23–63.

11. Souma W., Fujiwara Y., & Aoyama H. 2006. Heterogeneous economic networks. in: Namatame A., Kaizouji T., & Aruka Y. (eds). *The Complex Networks of Economic Interactions.* Lecture Notes in Economics and Mathematical Systems. Germany: Springer, 567, 79–92.

12. Battiston S., Rodrigues J.F., & Zeytinoglu H. 2007. The network of inter-regional direct investment stocks across. *Advances in Complex Systems*, 10(1), 29–51.

13. Glattfelder J.B., & Battiston S. 2009. Backbone of complex networks of corporations: The flow of control. *Physical Review E*, 80(3), 036104. https://doi.org/10.1103/PhysRevE.80.036104

14. Nakano T., & White D. 2007. Network structures in industrial pricing: The effect of emergent roles in Tokyo supplier-chain hierarchies. *Structures and Dynamics*, 2(3), 130–154.

15. Reyes J., Schiavo S., & Fagiolo G. 2008. Assessing the evolution of international economic integration using random-walk betweenness centrality: The cases of East Asia and Latin America. *Advances in Complex Systems*, 11(5), 685–702.

16. Lublóy A. 2006. Topology of the Hungarian large-value transfer system. *Magyar Nemzeti Bank (Central Bank of Hungary) MNB Occasional Papers*, 57.

17. Inaoka H., Nimoniya T., Taniguchi K., Shimizu T., & Takayasu H. 2004. Fractal network derived from banking transactions—An analysis of network structures formed by financial institutions. *Bank of Japan Working Papers Series*, Bank of Japan, 04(4).

18. Soramäki K., Bech M.L., Arnold J., Glass R.J., & Beyeler W.E. 2007. The topology of interbank payment flows. *Physica A*, 379(1), 317–333.

19. Boss M., Helsinger H., Summer M., & Thurner S. 2004. The network topology of the interbank market. *Quantitative Finance*, 4(6), 677–684.

20. Iori G., & Jafarey S. 2001. Criticality in a model of banking crisis. *Physica A*, 299(1), 205–212.

21. Iori G., De Masi G., Precup O.V., Gabbi G., & Caldarelli G. 2007. A network analysis of the Italian overnight money market. *Journal of Economic Dynamics and Control*, 32(1), 259–278.

22. Rendón de la T.S., Kalda J., Kitt R., & Engelbrecht J. 2016. On the topologic structure of economic complex networks: Empirical evidence from large scale payment network of Estonia. *Chaos Solitons and Fractals*, 90, 18–27.

23. Rendón de la T.S., Kalda J., Kitt R., & Engelbrecht J. 2017. Detecting overlapping community structure: Estonian network of payments (submitted).

24. Rendón de la T.S., Kalda J., Kitt R., & Engelbrecht J. 2017. Fractal and multifractal analysis of complex networks: Estonian network of payments, *European Journal of Physics B*, 90(12), 234–241.

25. West D.B. 2003. *Introduction to Graph Theory*. Upper Saddle River, NJ: Prentice Hall.

26. Erdös P., & Rényi A. 1959. On random graphs. *Publicationes Mathematicae Debrecen*, 6, 290–297.

27. Mandelbrot B. 1983. *The Fractal Geometry of Nature*. New York, NY: Academic Press.

28. Jeong H., Tombor B., Reka A., Zoltan N.O., & Barabási A.L. 2000. The large-scale organization of metabolic networks. *Nature*, 407, 651–654.

29. Barabási A.L., & Zoltan N.O. 2004. Network biology: Understanding the cell's functional organization. *Nature Review Genetics*, 5(2), 101–103.

30. Morone F., & Makse H.A. 2015. Influence maximization in complex networks through optimal percolation. *Nature*, 524(7563), 65–68.

31. Palla G., Derényi I., Farkas I., & Vicsek T. 2005. Uncovering the overlapping community structure of complex networks in nature and society. *Nature*, 435, 814–818.

32. Derényi I., Palla G., & Vicsek T. 2005. Clique percolation in random networks. *Physics Review Letters*, 94(16), 160–202.

33. Traud A.L., Mucha J.P., & Porter M.A. 2012. Social structure of Facebook networks. *Physica A*, 391(16), 4165–4180.

34. González M.C., Herrmann H.J., Kertész J., & Vicsek T. 2007. Community structure and ethnic preferences in school friendship networks. *Physica A*, 379(1), 307–316.

35. Palla G., Barabási A.L., & Vicsek T. 2007. Quantifying social group evolution. *Nature*, 446(7136), 664–667.

36. Pollner P., Palla G., & Vicsek T. 2006. Preferential attachment of communities: The same principle, but a higher level. *Europhyics Letters*, 73(3), 478–484.

37. Lewis A.C.F., Jones N.S., Porter M.A., & Deane C.M. 2010. The function of communities in protein interaction networks at multiple scales. *BMC Systems Biology*, 4, 100.

38. Guimerá R., & Amaral N. 2005. Functional cartography of complex metabolic networks. *Nature*, 433, 895–900.

39. Dourisboure Y., Geraci F., & Pellegrini M. 2007. Extraction and classification of dense communities in the web. *Proceedings of the 16th International Conference on the World Wide Web*, 1, 461–470.

40. Vitali S., & Battiston B. 2014. The community structure of the global corporate network. *Plos ONE*, 9(8), 104655. https://doi.org/10.1371/journal.pone.0104655

41. Fenn D. et al. 2009. Dynamic communities in multichannel data: An application to the foreign exchange market during the 2007–2008 credit crisis. *Chaos*, 19(3), 033119.

42. Piccardi C., Calatroni L., & Bertoni F. 2010. Communities in Italian corporate networks. *Physica A*, 389(22), 5247–5258.

43. Bóta A., & Krész M. 2015. A high resolution clique-based overlapping community detection algorithm for small-world networks. *Informatica*, 39, 177–187.

44. Yang Z., Algesheimer R., & Tessone C.J. 2016. A comparative analysis of community detection algorithms on artificial networks. *Scientific Reports*, 6, 30750.

45. Devi J.C., & Poovammal E. 2016. An analysis of overlapping community detection algorithms in social networks. *Proceedia Computer Sciences*, 89, 349–358.

46. Everett M.G., & Borgatti S.P. 1998. Analyzing clique overlap. *Connections*, 21(1), 49–61.

47. Shen H.W. 2013. *Community Structure of Complex Networks*. Germany: Springer Science & Business Media.

48. Newman M.E.J. 2004. Fast algorithm for detecting community structure in networks. *Phyical Review E*, 69, 066133.

49. Clauset A., Newman M.E.J., & Moore C. 2004. Finding community structure in very large networks. *Physical Review E*, 70(6), 066111. http://dx.doi.org/10.1103/PhysRevE.70.066111

50. Lu Y., & Tang J. 2004. Fractal dimension of a transportation network and its relationship with urban growth: A study of the Dallas-Fort Worth area. *Environmental and Planning B*, 31(6), 895–911.

51. Fotios M., & Siokis M. 2014. European economies in crisis: A multifractal analysis of disruptive economic events and the effects of financial assistance. *Physica A*, 395, 283–292.

52. Pont O., Turiel A., & Pérez-Vicente C.J. 2009. Empirical evidences of a common multifractal signature in economic, biological and physical systems. *Physica A*, 388(10), 2025–2035.

53. Song Y.Q., Liu J.L., Yu Z.G., & Li B.G. 2015. Multifractal analysis of weighted networks by a modified sandbox algorithm. *Scientific Reports*, 5, 17628.

54. Zhou W.X., Yiang Z.Q., & Sornette D. 2007. Exploring self-similarity of complex cellular networks: The edge-covering method with simulated annealing and log-periodic sampling. *Physica A*, 375(2), 741–752.

55. Gallos L.K., Song C., Havlin S., & Makse H.A. 2007. A review of fractality and self-similarity in complex networks. *Physica A*, 386, 686–691.

56. Schneider C.M., Kesselring T.A., Andrade J.S., & Herrmann H.J. 2012. Box-covering algorithm for fractal dimension of complex networks. *Physical Review E*, 86, 016707.

57. Eguiluz V.M., Hernández-García E., Piro O., & Klemm K. 2003. Effective dimensions and percolation in hierarchically structured scale-free networks. *Physical Review E*, 68(5), 551021–551024.

58. Song C., Gallos L.K., Havlin S., & Makse H.A. 2007. How to calculate the fractal dimension of a complex network: The box covering algorithm. *Journal of Statistical Mechanics: Theory and Experiment*, 3, P03006.

59. Song C., Havlin S., & Makse H.A. 2005. Self-similarity of complex networks. *Nature*, 433, 392–395.

60. Li B.G., Yu Z.G., & Zhou Y. 2014. Fractal and multifractal properties of a family of fractal networks. *Journal of Statistical Mechanics: Theory and Experiment*, 2, P02020.

61. Liu J.L., Yu Z.G., & Anh V. 2015. Determination of multifractal dimensions of complex networks by means of the sandbox algorithm. *Chaos*, 25, 023103.

62. Wei D.J. et al. 2013. Box-covering algorithm for fractal dimension of weighted networks. *Scientific Reports*, 3, 3049.

63. Wang D.L., Yu Z.G., & Anh V. 2012. Multifractal analysis of complex networks. *Chinese Physics B*, 21(8), 080504.

64. Feder J. 1988. *Fractals*. New York, NY: Plenum.

65. Kim J.S., Goh K.I., Salvi G., Oh E., Kahng B., & Kim D. 2007. Fractality in complex networks: Critical and supercritical skeletons. *Physical Review E*, 75, 177.

66. Grady D., Thiemann C., & Brockmann D. 2012. Robust classification of salient links in complex networks. *Nature Communication*, 3(864), 864.

67. Kim D.H., Noh J.D., & Jeong H. 2004. Scale-free trees: The skeletons of complex networks. *Physics Review E*, 70, 046126.

68. Goh K.I., Salvi G., Kahng B., & Kim D. 2006. Skeleton and fractal scaling in complex networks. *Physics Review Letters*, 96(1), 018701.

69. Gibbons A. 1985. *Algorithmic Graph Theory*. Cambridge, UK: Cambridge University Press.

70. Tél T., Fülöp A., & Vicsek T. 1989. Determination of fractal dimensions for geometrical multifractals. *Physica A*, 159(2), 155–166.

71. Halsey T.C., Jensen M.H., Kadanoff L.P., Procaccia I., & Shraiman B.I. 1986. Fractal measures and their singularities: The characterization of strange sets. *Physics Review A*, 33(2), 1141–1151.

72. Grassberger P., & Procaccia I. 1983. Characterization of strange attractors. *Physics Review Letters*, 50(5), 346–349.

# Chapter 9

## Predicting Macroeconomic Variables Using Financial Networks Properties

Petre Caraiani

### CONTENTS

## 9.1  Introduction

In the latest years, financial networks have become a widely used tool to analyze different phenomena related to financial markets. While the initial research effort has been focused on uncovering the static structure and properties of financial networks (see References 10, 15, or 16), there is now a more pronounced focus on the use of financial networks to actually predict the future dynamics or the changes in the state of the financial markets.

A significant size of the research has been devoted to the use of state indices of the financial markets that can be a good approximation of the state of the market. A very good example in this sense are the papers by Kenett et al. [8]

or [9]. For example, Kenett et al. [9] have provided a methodology to construct an index of the state of the network based on financial networks using data on S&P500 with a sample between 1999 and 2010. The derived index, called the index cohesive force, was shown to be able to characterize the state of the market. Following study has rather focused on the dynamics of such index, see Kenett et al. [8], who extended the previous analysis by considering applications of the index cohesive force as well as of meta-correlations different financial markets in the world.

Another recent strand of research has focused on finding different ways to uncover structural properties in financial networks that can help predict the dynamics of the financial markets.

Caraiani [2] derived an entropy measure based on the correlation network of the components DOW Jones Industrial Average (DJIA) index using the singular value decomposition (SVD). The main result was that such an index has a predictive power with respect to the aggregate stock market dynamics as characterized by DJIA index. Some studies have confirmed this property for different markets. For example, Gu et al. [6] applied the approach to Shenzhen stock market and found similar predictive power of the index. They also found that the predictive power is affected by the window used for correlation as well as by structural breaks.

Another line of research (see Reference 13 or 11) focused on the properties of the matrices of correlations between assets of financial markets indices to find whether they can help predict the financial crisis. This direction has been extended in Reference 1, which employed network measures to study the interdependence between U.S. equity and commodity future markets.

More recently, Caraiani [3] showed that we can also use the local properties of correlation-based financial networks to derive indices that have a predictive power with respect to the aggregate dynamics of the stock market.

Although the previous research was able to find structural properties of the networks that can help construct indices with predictive properties relative to the dynamics of the aggregate stock market, there is almost no research with respect to investigating the ability of such properties to predict the dynamics of macroeconomic and financial variables. A notable exception is the recent study by Heiberger [7], who used machine-learning techniques to show that the recent recessions were predictable based on the analysis of stock networks.

The aim of this paper extends the previous study in Reference 2 by studying whether the singular value decomposition-based entropy (SvdEn) has any predictive power with respect to key macroeconomic and financial variables.

---

## 9.2   Methodology

In this section, we focus on the main tools used throughout the paper. The presentation follows the previous methods described in Reference 2 as well as in Reference 3.

## 9.2.1 Correlation networks of stocks

In order to construct correlation-based financial networks, we used standard correlation (Pearson's correlation). A comparison of the different methods to compute the correlation is done in Reference 9.

To construct the correlation matrix, we use

$$\rho_{i,j} = \frac{cov(r_i, r_j)}{\sigma_i \sigma_j} \tag{9.1}$$

Here $r_i$ is the return of the a stock $i$, while $\sigma_i$ is the standard deviation of the return of the same stock $i$. The return of a stock $i$ is computed in a standard way as the logarithmic different of the value of that stock, namely by

$$r_i(t) = log[P_i(t)] - log[P_i(t-1)] \tag{9.2}$$

where $r_i(t)$ is the return of the stock $i$ at time $t$, while $P_i(t)$ is the value of the same stock $i$ at moment $t$.

By applying this approach, we can obtain a matrix $R$ which has the dimensions $N \times N$, where $N$ is the number of stock used in constructing the correlation matrix.

In the end, we get a matrix $R$ of dimensions $N \times N$, with $N$ is the number of stock included. As in the previous contributions (see Reference 2) we set a threshold of 0.3 to filter the weak correlations out. Based on the derived correlations, we compute the distances between the nodes as 1 less the correlation. After doing this for all nodes and links, we finally obtain an adjacency matrix between the stocks which are the nodes in the network.

## 9.2.2 Singular value decomposition

In this section, we introduce the SVD. The SVD can be applied to any matrix. In the context of this paper, it will be applied to the matrix of correlations at a certain point in time.

In the general case, a matrix $A$, having the dimensions $m \times n$ can be decomposed with the help of the SVD as follows:

$$A = USV^T \tag{9.3}$$

In the above equation, the matrix $U$ is of dimensions $m \times k$ while $V$ is a matrix of dimensions $k \times n$. Accordingly, the matrix $S$ must be a diagonal matrix defined by:

$$S = \text{diag}(\lambda_1, \lambda_2, \ldots, \lambda_k) \tag{9.4}$$

The dimension $k$ is given by $k = \min(m,n)$. The matrix $S$ is constructed such that its elements have two essential characteristics: they are positive and they are ordered from the biggest to the lowest.

### 9.2.3 Singular value decomposition-based entropy

We now introduce a global measure of networks used throughout the paper, that is, the SvdEn. Based on the SVD method presented before, one can construct an entropy measure. This is based on the contribution of Sabatini [12], who actually extended the former work by Shannon [14].

To construct the SvdEn, one uses the singular values $\lambda_k$ extracted based on the SVD method, see the above section. The method consists of a few simpler steps. First, we compute

$$\lambda_k = \frac{\lambda_k}{\sum \lambda_k} \tag{9.5}$$

This formula implies that the singular values are normalized. To derive the entropy measure, denoted by $E$, the following formula is used:

$$E = -\sum \lambda_k ln(\lambda_k) \tag{9.6}$$

### 9.2.4 Granger causality tests

The Granger causality tests is meant to test whether a given series can help predict another series. This is due to the contributions by Granger [4,5]. We describe below the test in a succinct manner.

$$y_t = \beta_0 + \sum_{i=1}^{N} \beta_k y_{t-k} + \sum_{l=1}^{N} \alpha_l x_{t-l} + u_t \tag{9.7}$$

$y_t$ and $x_t$ are the two given time series, while $u_t$ represents the disturbances. The parameters $k$ and $l$ stand for the number of lags.

In the Granger causality test, the null hypothesis is that the series $y$ does not cause the series $x$. Using the representation above, this implies that there is no $a_l$ significantly different from zero. The null and alternative hypothesis are written below:

$H0 : \alpha_l = 0$ for any $l$;

$H1 : \alpha_l \neq 0$ for at least some $l$.

---

## 9.3 Results

### 9.3.1 Data selection

We follow previous papers by Caraiani [2,3] and select data on the main U.S. stock market index, the DJIA. The choice is based on a few arguments. First, the index has been used in previous papers that dealt with a similar topic (see again Reference 2). Second, since we use the components of DJIA, it is much easier to work with a limited number of stocks. Monthly observations were selected for DJIA series as well as its components. The data sample spans from July 1986 to July 2017. The main reason for selecting monthly observations was that the paper focuses focus on the relationship between stock market dynamics and macroeconomic data which are mainly available at monthly frequencies. Appendix 9A presents the stocks included in the sample as well as their acronyms. A few stocks whose sample was limited were excluded (namely Cisco Systems, Goldman Sachs, United Health Group, and Visa Inc.).

It must also be pointed out that the selected components were based on the current composition of DJIA. However, the limitation of the study is that the methodology makes it more difficult in changing the selected group of stocks.

We have also selected a number of macroeconomic and financial variables as follows: consumer price index (CPI, hereafter), Federal Funds Rate, the Industrial Production, money stock M2 (Nominal and Real, that is deflated by CPI), Unemployment, as well as 3 Month Treasury Bill. A similar data sample was selected for these variables too.

### 9.3.2    Correlation-based financial networks

In this section, we present the derived measure of entropy based on the selected stocks. Since this paper proposes a dynamic analysis, we derive a time-varying measure of entropy by constructing correlation networks among the selected stocks on a sliding window. We selected a sliding window of 24 observations (i.e., 24 months or 2 years). In each period, the sliding window is moved one observation to the right, the correlation network is constructed, and the corresponding SvdEn is derived. This approach ensures that we can construct a time series for SVDEn and analyze its relationship with the selected macroeconomic and financial variables.

### 9.3.3    The singular value decomposition-based entropy

The derived measures of SvdEn, both as raw (index) series and as growth series, are shown in Appendix 9B. Both the raw (index) and the growth series show significant variation. This has also been pointed out in previous similar studies, see Reference 2 which also studied SvdEn and Reference 3 which studied the variability of local properties. The index series also shows a significant spike in variation around the time of the financial crisis, suggesting that it might be correlated or that it might even have explanatory power for the variation in the dynamics of aggregate financial market. We study this issue more in-depth in the next section.

**TABLE 9.1:**    Granger causality test

| Variable | Index | Growth rate |
|---|---|---|
| DJIA | 5.72* | 3.67 |
| CPI | 0.13 | 2.41 |
| Federal Funds Rate | 6.10* | 3.14 |
| Industrial Production | 0.00 | 1.14 |
| M2 Nominal | 0.04 | 0.68 |
| M2 Real | 0.00 | 2.75 |
| 3 Month Treasury Bill | 4.51* | 5.30* |
| Unemployment | 3.96* | 3.50 |

*Note:* *Statistical significance of the F-test at 0.05 level.

### 9.3.4    Granger causality results

The analysis in the previous section suggested that the derived measure of SvdEn might be at least correlated with the dynamics of the aggregate stock market index. In this section, we analyze this issue by considering whether the two measures of entropy, the raw and the growth measures, have prediction power in terms of Granger causality with respect to macroeconomic and financial variables.

We present below the results of the Granger causality tests for the two measures of entropy with respect to the selected macroeconomic and financial variables. The lag order in the Granger causality test is optimally selected from up to 12 lags using the Bayesian information criterion (also known as BIC).

Table 9.1 shows the results of the Granger causality tests. In most cases, it is the raw or index version of entropy that has a prediction power over some of the macroeconomic and financial variables, namely DJIA, Federal Funds Rate, 3 Month Treasury Bill, and Unemployment. For the case of entropy in growth rates, this has a prediction power only for the 3 Month Treasury Bill.

### 9.3.5    Further statistical evidences

We extend in this section, the previous analysis by considering whether the two measures of entropy have any explanatory power for the variation of the selected macroeconomic and financial variables. To test this, we use the following regression specification:

$$y_t = \alpha + \beta * SvdEn_t + \gamma * DUMMY\,2001 + \delta * DUMMY\,2008 + \epsilon_t \quad (9.8)$$

where $y_t$ is the value of the macroeconomic or financial variable at time $t$, $\alpha$ is a constant, $SvdEn$ is the value of the SvdEn, $DUMMY\,2001$ and $DUMMY\,2008$ stand for the dummy variables that capture the large movements due to financial crises while $\epsilon_t$ is the residual.

**TABLE 9.2:**  Regression analysis I

| Dependent | log(DOW) | log(CPI) | Federal funds rate | Industrial production |
|-----------|----------|----------|--------------------|-----------------------|
| $\alpha$ | 9.17*** | 5.13*** | 3.05*** | 4.24*** |
| SvdEn | 0.15*** | 0.00 | 0.09 | 0.01*** |
| *DUMMY* 2001 | −0.07** | 0.003 | 0.04 | −0.02 |
| *DUMMY* 2008 | 0.02 | 0.004** | 0.30 | −0.03 |
| $R^2$ | 0.99 | 0.99 | 0.73 | 0.74 |

*Note:* *Statistical significance at 0.1 level; **Statistical significance at 0.05 level; ***Statistical significance at 0.01 level.

**TABLE 9.3:**  Regression analysis II

| Dependent | M2 | M2 real | Unemployment | 3 month treasury bill |
|-----------|-----|---------|--------------|-----------------------|
| $\alpha$ | 8.90*** | 8.23*** | 12.17*** | 3.30*** |
| SvdEn | −0.09 | −0.006*** | −2.50*** | 0.26* |
| *DUMMY* 2001 | 0.29 | −0.007*** | 0.20 | −0.06 |
| *DUMMY* 2008 | 0.44 | −0.0009 | 2.28 | 0.01 |
| $R^2$ | 0.00 | 0.99 | 0.21 | 0.99 |

*Note:* *Statistical significance at 0.1 level; **Statistical significance at 0.05 level; ***Statistical significance at 0.01 level.

Tables 9.2 and 9.3 present the results of the regression analysis. Since the raw or index version of the derived measure of entropy has more predictive power, we focused only in using this measure. The results basically confirm the findings from the Granger causality tests. The SvdEn significantly affects the DJIA variable, as well as the Industrial Production. It also affects the real money stock M2, the Unemployment, as well as the 3 Month Treasury Bill.

### 9.3.6   Discussion of results

The above results found clear evidences that a systemic measure of financial networks for U.S. stock market can help explain the dynamics of key macroeconomic and financial variables. In terms of predictive ability, evidences were weaker, underlining, however, the ability to predict the future dynamics of DJIA stock market index, Federal Funds Rate, Unemployment, or 3 Month Treasury Bill. For DJIA this is rather expected, since the financial networks were derived using the components of DJIA. It is interesting, however, to learn that the SvdEn was also able to predict the change in Unemployment or key interest rates. This underlines the fact that SvdEn reveals the state of the market, a fact revealed by previous similar studies (see Reference 2 or 9).

With respect to modeling the dynamics of the same selected macroeconomic and financial variables, as shown by the estimated regressions, the explanatory power of the SvdEn is significant across a wider range of variables. The regression analysis revealed that the SvdEn can help explain the variation in DJIA index, Industrial Production, Real M2 stock, Unemployment, or 3 Month Treasury Bill. Although the results differ relative to the findings based on the Granger causality tests, nevertheless, these are further evidences that this network-based entropy can have explanatory power with respect to the dynamics of the macroeconomy.

## 9.4    Conclusion

Most of the previous papers studied the potential predictive and explanatory power of systemic measure of entropy of financial networks or local measures of financial networks with respect to the aggregate stock market dynamics. The results have been, in general, positive as the systemic or local measures of networks do help predict or detect the changes in the state of the financial market.

However, not much work has been done with respect to the role of systemic measures (or local measures) to predict the dynamics of key macroeconomic and financial variables. In this paper, we addressed this issue by looking at the predictive and explanatory power of the entropy of financial networks with respect to several macroeconomic and financial variables of interest. The results show that indeed the SvdEn of financial networks does have predictive power in terms of Granger causality and explanatory power in terms of regression analysis with respect to the variation in selected macroeconomic and financial variables.

The results here encourage further analysis of the interdependence between financial networks and measures of economic activity. Future analysis could take into account networks that arise naturally in economics and finance, such as trade networks, banking and financial networks, or production networks, and extend the tools here to analyze the information content in their variation with respect to aggregate dynamics of the macroeoconomy or financial markets. Some promising venue of research might come from offering an integrated view of the macroeconomy and financial markets based on networks.

# Appendix 9A  Data Series Components for DJIA Index

**TABLE 9A.1:**   DOW jones industrial average components

| Company | Abbreviation |
| --- | --- |
| Apple Inc. | AA |
| American Express Company | AXP |
| Boeing Company | BA |
| Caterpillar, Inc. | CAT |
| E.I. du Pont de Nemours and Com | DD |
| Chevron Corporation | CVX |
| Walt Disney Company (The) | DIS |
| General Electric Company | GE |
| Home Depot, Inc. (The) | HD |
| International Business Machines | IBM |
| Intel Corporation | INTC |
| Johnson & Johnson | JNJ |
| JP Morgan Chase & Co. | JPM |
| Coca-Cola Company (The) | KO |
| McDonald's Corporation | MCD |
| 3M Company | MMM |
| Merck & Company, Inc. | MRK |
| Microsoft Corporation | MSFT |
| NIKE, Inc. | NKE |
| Pfizer, Inc. | PFE |
| Procter & Gamble Company (The) | PG |
| The Travelers Companies, Inc. | TRV |
| United Technologies Corporation | UTX |
| Verizon Communications Inc. | VZ |
| Wal-Mart Stores, Inc. | WMT |
| Exxon Mobil Corporation | XOM |

# Appendix 9B  The Time Series for the Derived SvdEn

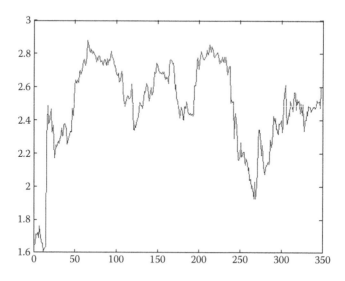

**FIGURE 9B.1:** SvdEn—Index series.

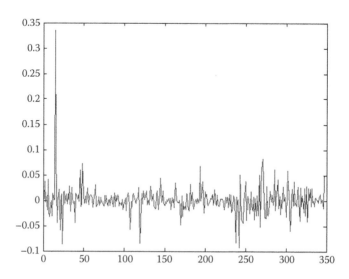

**FIGURE 9B.2:** SvdEn—Growth rate series.

# References

1. S. Bekiros, D. Khuong Nguyen, L. Sandoval Junior, and G. Salah Uddin. Information diffusion, cluster formation and entropy-based network dynamics in equity and commodity markets. *European Journal of Operational Research*, 256(3):945–961, 2017.

2. P. Caraiani. The predictive power of singular value decomposition entropy for stock market dynamics. *Physica A*, 393(1):571–578, 2014.

3. P. Caraiani. The predictive power of local properties of financial networks. *Physica A: Statistical Mechanics and its Applications*, 466(C):79–90, 2017.

4. C.W.J. Granger. Investigating causal relations by econometric models and cross-spectral methods. *Econometrica*, 37(3):424–438, 1969.

5. C.W.J. Granger. Testing for causality : A personal viewpoint. *Journal of Economic Dynamics and Control*, 2(1):329–352, 1980.

6. R. Gu, W. Xiong, and X. Li. Does the singular value decomposition entropy have predictive power for stock market? Evidence from the Shenzhen Stock Market. *Physica A*, 439(1):103–113, 2015.

7. R.H. Heiberger. Predicting economic growth with stock networks. *Physica A: Statistical Mechanics and its Applications*, 489:102–111, 2018.

8. D.Y. Kenett, M. Raddant, Th. Lux, and E. Ben-Jacob. Evolvement of uniformity and volatility in the stressed global financial village. *PLoS ONE*, 7(2):e31144, 2012.

9. D.Y. Kenett, Y. Shapira, A. Madi, S. Bransburg-Zabary, G. Gur-Gershgoren, and E. Ben-Jacob. Index cohesive force analysis reveals that the us market became prone to systemic collapses since 2002. *PLoS ONE*, 6(4):e19378, 2011.

10. R.N. Mantegna. Hierarchical structure in financial markets. *The European Physical Journal B*, 11:193–197, 1999.

11. K.D. Peron, L. da Fontoura Costa, and F.A. Rodrigues. The structure and resilience of financial market networks. *Chaos*, 22(1), 2012.

12. A.M. Sabatini. Analysis of postural sway using entropy measures of signal complexity. *Medical and Biological Engineering and Computing*, 38(6):617–624, 2000.

13. L. Sandoval Junior and I. De Paula Franca. Correlation of financial markets in times of crisis. *Physica A: Statistical Mechanics and its Applications*, 391(12):187–208, 2012.

14. C.E. Shannon. A mathematical theory of communication. *Bell System Technical Journal*, 27(3):379–423, 1948.

15. M. Tumminello, T. Aste, T. Di Matteo, and R. N. Mantegna. A tool for filtering information in complex systems. *Proceedings of the National Academy of Sciences of the United States of America*, 102:10421–10426, 2005.

16. M. Tumminello, T. Di Matteo, T. Aste, and R.N. Mantegna. Correlation based networks of equity returns sampled at different time horizons. *The European Physical Journal B*, 55:209–217, 2007.

# Chapter 10

## Anomaly Detection in Dynamic Complex Networks

Yasser Yasami

### CONTENTS

## 10.1   Introduction

Nowadays, social networks are applied in various domains such as education, business, and many others. The social networks are often dynamic and relationships between social entities vary over time. One of the fundamental issues of such dynamic networks is to model the dynamics of the underlying relationships between entities and extract a summary of the common normal structure. The extracted models of dynamic social networks can be useful for missing links prediction [1–6], potential link recommendation [7], community detection [8,9], future relationships forecasting [10–13], group growth and longevity prediction [14], as well as anomaly detection. Besides, the widespread application of social networks comes with increasing the prevalence of malicious activities using social networks. Accordingly, the development of methods for detecting anomalous entities in dynamic social networks is critical to coincide with the growth in social networks usage [15].

Detecting anomalous entities in social networks refers to detecting existing links/nodes that are not expected to exist as well as non-existing links/nodes that are expected to exist. In this sense, the social networks anomaly detection not only bears some similarities to link prediction, link recommendation, and missing link inferences, but it also encompasses these concepts.

While detecting anomalous entities in social networks has been addressed extensively in the literature, previous works mostly ignore to consider the dynamics of microscopic features of network entities. However, this chapter addresses the problem of detecting anomalous entities in dynamic social networks by considering dynamic infinite features of network entities and dynamic feature cascading. To cope with this problem a statistical approach for modeling of dynamic social networks is developed, that is capable of modeling the cascade behavior of infinite features, also referred to as social influence. The proposed approach can identify the birth, death, and longevity of individual features as well as networks entities taking and giving up the features in order to model dynamics of network structure. The proposed approach considers an infinite number of features, where each network entity can take multiple features, simultaneously. The dynamics of each individual entity feature is extracted by a Factorial Hidden Markov model (FHMM).

One of the main contributions of this chapter is considering features cascade in modeling of dynamic social networks. The other contribution is that in contrast to recent related works [10,12,16–18], in order to avoid some ambiguities, this chapter explicitly model the birth and death dynamics of individual features by an Indian Buffet Process (IBP) [19], under which the longevity of features is assumed to follow a more realistic distribution; an exponential distribution. Under the proposed model, only active features influence the relationships between entities in a network at a given time step. Further contribution of this chapter is that the relationships between entities with different features are also considered in addition to the ones between entities with the same features. Moreover, this chapter extends the anomaly detection approach proposed in [44] to detect anomalous nodes. The proposed approach is validated through a number of statistical measures of the performance specially applied in the normal/abnormal binary classification test. By explicit modeling of feature birth, longevity and death, feature cascade, and feature connectivity structure, greater modeling flexibility is achieved, which leads to improved performance on detecting anomalous links and anomalous nodes based on the defined criteria as well as to increased interpretability of obtained results.

After this introduction, the rest of this chapter is organized as follows: Section 10.2 reviews the background and related work of this research field.

Section 10.3 briefly describes the general modeling of dynamic social networks. The proposed infinite feature cascade model is provided in Section 10.4. Section 10.5 describes how to detect anomalous entities, including anomalous links and anomalous nodes in dynamic social networks. Section 10.6 includes model inference using Markov Chain Monte Carlo (MCMC), experimental results and analysis of real dynamic social networks. Finally, Section 10.7 contains conclusions, discussion, and potential future works.

## 10.2   Background and Related Work

Anomaly detection in complex networks is a vibrant research area and this problem has been addressed in many literatures. Authors in Reference 20 present a two-stage anomaly detection approach in dynamic graphs. The first stage uses statistical models for discrete time counting processes to track the pairwise links of all nodes in the graph to assess normality of behavior. The second stage applies standard network inference tools on a greatly reduced subset of potentially anomalous nodes. Authors in Reference 21 propose a framework for analyzing the effectiveness of various macroscopic graph theoretic properties. A community-based anomaly detection for social networks has been proposed in Reference 22. All of these approaches are unable to consider microscopic features of network entities for anomaly detection task.

Modeling the normal behavior of dynamic social networks is one of the main components of the proposed anomaly detection approach. There are non-Bayesian approaches [23,24] that have been used to model static as well as dynamic social networks. Some of these approaches are based on Exponential-family Random Graph Models that can be represented via a particular canonical parameterization [23]. These methods, however, suffer from both modeling framework inconsistencies and computational difficulties [25].

Some of other non-Bayesian approaches [26,27] for modeling dynamic network structure as dependent on the observed structure at previous time steps are usually referred to as autoregressive models that typically deal with only the networks observations and do not consider latent variable representations. Furthermore, cascade phenomenon is studied in some social networks literatures, often termed as social influence, selection, and trust propagation [28]. Some of the collaborations do use latent variable approaches, yet lack a probabilistic framework.

Nevertheless, some Bayesian approaches to dynamic network analysis have been proposed in which latent variable models have been used. There are differences among these approaches, mostly in the structure of the latent space that they apply. For example, the entities in Euclidean space models [13,29] are assumed to be in a low-dimensional Euclidean space. Then, the network evolution is modeled as a regression problem of the entity's future latent location. Other latent variable representations of dynamic networks have been presented in References 16 and 30, where the evolution of the network through time is determined by the underlying latent variables, evolving according to a linear Gaussian model. Latent representations are also applied in Reference 31 and a particular study of social networks in Reference 29, in which the network evolution is considered as a regression problem, where the parameters of the model represent expectation and covariance values of the networks connectivity patterns.

Related to the work presented in this chapter are dynamic mixed-membership models where a node is probabilistically allocated to a set of latent features, including the dynamic mixed-membership block model [16,17] and the dynamic infinite relational model [18]. However, in contrast to the approaches presented in References 16–18, the proposed approach applies infinite multi-features model where each network entity's feature does not limit that entity to possess other features.

The proposed model applies the ideas of the recently introduced infinite FHMM [32], an approach that modifies the IBP into an FHMM with an unbounded number of hidden chains. Modeling temporal changes in latent variables, for entities in a network, has been also proposed in References 13, 16 and 33. A major difference is that the proposed approach models an entity's evolution by Markov switching rather than via the Gaussian linear motion models used in these papers. The proposed approach explicitly models the dynamics of the entities' latent representations, where unlike the model proposed in Reference 16, making it more suitable for anomaly detection.

Probably, highly related to the work presented in this chapter are DRIFT [10], LFP [12], and DMMG [34] models. All these models consider Markov switching of latent multi-group memberships over time. DRIFT and DMMG uses the infinite FHMM [32], while LFP adds "social propagation" to the Markov processes so that network links of each node at a given time directly influence group memberships of the corresponding node at the next time. All these models have shortcomings. For example, DRIFT which is the dynamic extension of the latent feature relational model (LFRM) [2], and DMMG are unable to model social cascade. Furthermore, DMMG assumes features longevities to follow a geometric distribution over discrete time steps, which is not a realistic distribution for features of dynamic social networks human entities. LFP is also faced with some ambiguities because it is unable to explicitly model birth, death, and longevity of features. On the other hand, the proposed statistical infinite feature cascade model uniquely incorporates the birth, longevity and death model of features, and features cascading. The performance of the proposed approach is presented through statistical measures of performance of the binary classification test of normal/abnormal over some real dynamic social networks.

To the best of our knowledge, the proposed model and anomaly detection approach is the first statistical infinite feature cascade approach for dynamic social networks which has higher performance in comparison with other related approaches.

## 10.3   Modeling of Dynamic Social Networks

This section introduces the general components and notations used for modeling dynamic social networks as applied in other related literatures [10,12,16,18]. Table 10.1 summarizes the notations and symbols used in this chapter and their description.

Dynamic networks can be generally represented as discrete time series of graphs $G^{(t)} = (N^{(t)}, Y^{(t)})$, $t = 1, 2, \ldots, T$. A set of nodes $N^{(t)}$ represents a set of people at time $t$ and a network connectivity matrix $Y^{(t)}$ that is represented as an adjacency matrices, interprets different relationships among people such as friendship, affiliation, family, and many others at time $t$. Such a model can be applied far more general in many branches of physics, biology, etc. This chapter is only focused on unweighted, directed links and does not consider self-links. Therefore, each element of $Y^{(t)}$, an $|N^{(t)}| \times |N^{(t)}|$ binary matrix, where $|N^{(t)}|$ is the number of nodes at time step $t$, can be described as:

$$Y^{(t)} = (y_{ij}^{(t)}) \in \{0, 1\}^{|N^{(t)}| \times |N^{(t)}|},$$

$$y_{ij}^{(t)} = \begin{cases} 1, & \text{if there is a link from node } i \text{ to node } j \text{ at time } t, \\ 0, & \text{otherwise.} \end{cases} \qquad (10.1)$$

In modeling dynamic social networks, three main processes need to be explicitly described. The first process is related to the dynamics of nodes features. A vector of binary features is associated with each node $i$ of the network. This feature vector is considered to influence the interaction dynamics of the network nodes with each other. The binary value of feature $k$ of node $i$ at time $t$ is denoted by $z_{ik}^{(t)}$ and is formally described as:

$$z_{ik}^{(t)} \in \{0, 1\}, \quad 1 \leq i \leq N^{(t)}, \quad k \in \mathbb{N}. \qquad (10.2)$$

Such features can be regarded as nodes microscopic features or assigning nodes to multiple overlapping, clusters or groups such as social communities of people with the same interests or hobbies [2,8]. The model proposed in this chapter make it possible for each node to have multiple features, simultaneously. Furthermore, each node feature is modeled using a separate Bernoulli random variable [9,35,36]. One of the main merits of this model is that the network nodes under this model are not limited to have other features, too. This is in contrast to mixed-membership models in which the distribution over individual node's features is modeled using a multinomial distribution [8,16,17].

The second process governing dynamics of social networks is a process under which nodes dynamically take and give up features. It can be assumed that each node $i$ can take or give up each feature $k$ according to a Markov model, in which feature cascade has been taken into account. However, as each node can take multiple features independently, naturally FHMM [32] is developed, in which

**TABLE 10.1:**   Table of notations and symbols

| Symbol | Description | Symbol | Description |
|---|---|---|---|
| $G^{(t)}$ | Dynamic network at time step $t$ | $\eta_i$ | Node $i$ susceptibility of being influenced by its neighboring nodes |
| $N^{(t)}$ | Set of nodes in the dynamic network at time step $t$ | $\omega_j$ | Influence weight of the neighboring node $j$ from the neighbor set of node $i$ |
| $Y^{(t)} = (y_{ij}^{(t)}) \in \{0, 1\}^{|N^{(t)}| \times |N^{(t)}|}$ | Dynamic network observation (network connectivity matrix) at time step $t$ | $\varphi(i, t)$ | Set of neighboring nodes of node $i$ |
| $z_{ik}^{(t)}$ | Binary value of feature $k$ of node $i$ at time step $t$ | $\theta_k$ | Link affinity matrix of feature $k$ |
| $Q_k^{(t)} = (q_k^{(t)}) \in [0, 1]^{2 \times 2}$ | Markov transition probability matrix of feature evolution at time step $t$ | $p_{ij}^{(t)}$ | Probability of generating link between nodes $i$ and $j$ at time step $t$ |
| $t_k^a$ | Number of time steps in which feature $k$ is active | $g(z_{i\cdot}^{(t)}, z_{j\cdot}^{(t)})$ | Link probability function |
| $t_k^d$ | Death time step of feature $k$ | $\beta_t$ | Bias parameter |

*(Continued)*

**TABLE 10.1 (Continued):** Table of notations and symbols

| Symbol | Description | Symbol | Description |
|---|---|---|---|
| $t_k^b$ | Birth time step of feature $k$ | $A^{(t)} = (a_{ij}^{(t)}) \in [0,1]^{|N^{(t)}| \times |N^{(t)}|}$ | Stochastic anomaly matrix at time step $t$ |
| $\kappa_t$ | Set of active features at time step $t$ | $LAS^{(t)} = (las_{ij}^{(t)}) \in \{0,1\}^{|N^{(t)}| \times |N^{(t)}|}$ | Link anomaly score matrix at time step $t$ |
| $K_t$ | Number of active features at time step $t$ | $NAS^{(t)} = (nas_i^{(t)}) \in \{0,1\}^{|N^{(t)}|}$ | Node anomaly score matrix at time step $t$ |
| $\gamma$ | Control hyperparameter of feature birth and death rates | $LR+$ | Positive likelihood ratio |
| $W_k^{(t)}$ | State (active/inactive) of feature $k$ at time step $t$ | $LR-$ | Negative likelihood ratio |
| $\kappa_t^+$ | Set of newly activated features at time step $t$ | $PV+$ | Positive predictive value |
| $K_t^+$ | Number of newly activated features at time step $t$ | $PV-$ | Negative predictive value |

features of each node evolve over time, independently. On the other statement, the evolution of each node feature $z_{ik}^{(t)}$ is driven through a $2 \times 2$ Markov transition probability matrix $Q_k^{(t)}$, where each entry $Q_k^{(t)}[r, s]$ could be defined by the following conditional probability [12,32]:

$$Q_k^{(t)} = (q_k^{(t)}) \in [0, 1]^{2 \times 2},$$

$$Q_k^{(t-1,t)}[r, s] = P(z_{ik}^{(t)} = s | z_{ik}^{(t-1)} = r, Y^{(t-1)}), \quad r, s \in \{0, 1\}, \quad (10.3)$$

in which "0" and "1" indicate non-member and member, respectively.

The third process is the one governing on the relationship between node features and the network links. Given node feature vectors $z_{i.}^{(t)}$ and $z_{j.}^{(t)}$ at time $t$, then the probability of linking between nodes $i$ and $j$ can be determined by the link the function $f$, formally described as follow:

$$Y^{(t+1)} \propto f(z_{i.}^{(t)}, z_{j.}^{(t)}, Y^{(t)}). \quad (10.4)$$

Detailed description about each of the mentioned processes of dynamic social networks is provided in the next section.

## 10.4     The Infinite Feature Cascade Model of Dynamic Networks

This section provides a full description of the proposed infinite feature cascade model of dynamic networks. The model is involved in three main processes governing network dynamics. The first process is related to feature birth, feature longevity, and feature death. The second process is the one governing on the evolution of node features. The third one explains relationship between node features and network edges. The following subsections describe all of these processes, extensively.

### 10.4.1     Birth, longevity, and death process of microscopic features in dynamic networks

Dynamics of networks are influenced by each node microscopic features, as well as feature birth, longevity, and death processes. As the example depicted in Figure 10.1 shows, each feature $k$, $k = 1, 2, \ldots, 5$, in a dynamic network can be born at time step $t_k^b$, be active for a specific period of time $t_k^a$, and finally die after this activation period forever at time step $t_k^d$, where:

$$t_k^a = t_k^d - t_k^b,$$

$$t_k^b \le t_k^d. \quad (10.5)$$

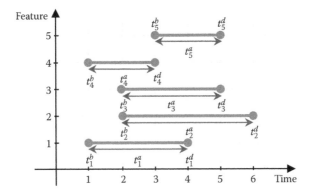

**FIGURE 10.1:** A graphical representation of feature birth, longevity, and death. (Adapted from Y. Yasami & F. Safaei, 2017. *Computer Communications*, 100, 52–64. [44])

Without explicit modeling of the feature birth, longevity, and death process, the modeling of network dynamics is really impossible, because of ambiguities between nodes features, birth, and death time steps of features. To avoid these ambiguities a model is proposed based on the idea of IBP [19,37], a probability distribution on sparse binary matrices with a finite number of rows, but an infinite number of columns.

The IBP is developed after a metaphorical process, where $N$ customers enter an Indian buffet restaurant and sample some subset of an infinitely long sequence of dishes. This process gives rise to the probability distribution in which the first customer samples the first $Poisson(\lambda)$ dishes, and the $k$th customer then samples the previously sampled dishes proportionately to their popularity, and samples $Poisson(\lambda/k)$ new dishes. A typical application of the IBP is to use it as a prior on a matrix that specifies the presence/absence of latent features that explain some observed data as used in References 2 and 38.

However, this chapter suggests a somewhat different approach. Each time step $t$ is regarded as a "customer," who samples a set of active features $\kappa_t$, where for initial state of $t = 1$, we have

$$|\kappa_1| \sim Poisson(\lambda), \tag{10.6}$$

where $|\kappa_1|$ denotes the number of active features at the first time step. In order to take the feature death, it is considered that each active feature at time $t - 1$ may die at the next time step $t$. To control for the rate of feature birth and death, a hyperparameter $\gamma$ is introduced. As $\gamma$ decreases, the rates of feature birth/death will decrease, while with an increase in $\gamma$, these rates increase, too. Thus, we have:

$$|\kappa_t^+| \sim Poisson(\gamma\lambda),\ t > 1, \tag{10.7}$$

where, $|\kappa_t^+|$ is the number of newly activated features at time step $t$ that is formally described later in this subsection. The above equation states that *Poisson* $(\gamma\lambda)$ new features are born at time $t$. Therefore, at each time step, currently active features may die, while the new ones can also be born.

Now, it is possible to define the probability that an active feature $k$ at time step $t-1$ remains active for the next time step $t$ as follows:

$$P(W_k^{(t)}|W_k^{(t-1)} = 1) = (1-\gamma)e^{(\gamma-1)(t-t_k^b)}, \qquad (10.8)$$

where, $W_k^{(t)}$ in the above equation is the state of feature $k$, at time step $t$, that is, active/inactive:

$$W_k^{(t)} = 1, \quad \{k \in \kappa_t\}. \qquad (10.9)$$

As it is obvious from Equation 10.8, feature longevity is considered to follow the exponential distribution, where

$$P(t_k^a \geq T) = e^{(\gamma-1)T}. \qquad (10.10)$$

This is in contrast to the model proposed in Reference 34, where describes the feature longevity by a nonrealistic distribution. Authors in Reference 34 applied a geometric distribution for the feature longevity, while this chapter considers an exponential distribution instead. The reason is that node features in dynamic social networks rise due to human collective behavior. We believe that the longevity of such features cannot really follow a geometric distribution over discrete time steps. As shown in Section 10.6, the experimental results indicate that the application of exponential distribution for modeling the feature longevity leads to improved performance in terms of anomaly detection characteristics compared to the model proposed in Reference 34. This means that the proposed model is a more realistic approach for modeling the feature longevity. In order to describe this process formally, the number of active features at time $t$ is denoted by $K_t$, where

$$K_t = |\kappa_t|. \qquad (10.11)$$

Moreover, a set of newly activated features and the number of newly activated features at time $t$ are denoted by $\kappa_t^+$ and $K_t^+$, respectively, which can be formally described as:

$$\kappa_t^+ = \{k|W_k^{(t)} = 1, W_k^{(t')} = 0, \quad \forall t' < t\}, \qquad (10.12)$$

$$K_t^+ = |\kappa_t^+|. \qquad (10.13)$$

Finally, the process of feature birth, longevity, and death that is graphically represented in Figure 10.1, can be formally described. The number of newly activated features at time step $t$, $K_t^+$, can be formally expressed as follows:

$$K_t^+ \sim \begin{cases} Poisson(\lambda), & \text{if } t = 1 \\ Poisson(\gamma\lambda), & \text{if } t > 1 \end{cases} \tag{10.14}$$

and the feature state at time step $t$, $W_k^{(t)}$, is defined as follows:

$$W_k^{(t)} \sim \begin{cases} Bernouli((1-\gamma)e^{(\gamma-1)(t-t_k^b)}), & \text{if } W_k^{(t-1)} = 1, \\ 1, & \text{if } \sum_{i=1}^{t-1} K_i^+ < k \le \sum_{i=1}^{t} K_i^+, \\ 0, & \text{Otherwise.} \end{cases} \tag{10.15}$$

## 10.4.2 Cascade modeling for microscopic features of social networks

Nodes in social networks can be influenced by each other. On the other hand, microscopic features in social networks can cascade from one node to other neighboring nodes as the network evolves over time. This means that each person can influence his friends.

In order to model dynamics of node features, an FHMM is applied that takes into account feature cascades as:

$$Q_k^{(t-1,t)}[r, s] = P\left(z_{ik}^{(t)} = s | z_{ik}^{(t-1)} = r, Y^{(t-1)}\right) = \left[\rho_{ik}^{(t)}\right]^s \left[1 - \rho_{ik}^{(t)}\right]^{(1-s)}, \tag{10.16}$$
$$r, s \in \{0, 1\},$$

where

$$\rho_{ik}^{(t)} = f_k\left(\mu_{ik}^{(t)}\right), \tag{10.17}$$

$$f_k(x) = \frac{c_k}{1 + e^{-x}} + b_k \tag{10.18}$$

and

$$\mu_{ik}^{(t)} = \frac{\eta_i \left[z_{ik}^{(t-1)} + \sum_{j \in \phi(i,t-1)} \omega_j z_{jk}^{(t-1)}\right]}{1 + \sum_{j' \in \phi(i,t-1)} \omega'_j} + (1 - \eta_i) z_{ik}^{(t-1)}. \tag{10.19}$$

However, if $Q_k^{(t-1,t)}$ is defined as:

$$Q_k^{(t-1,t)} = \begin{bmatrix} q_{k_{11}} & q_{k_{12}} \\ q_{k_{21}} & q_{k_{22}} \end{bmatrix}, \tag{10.20}$$

then $z_{ik}^{(t)}$ can be calculated by the following equation:

$$P\left(z_{ik}^{(t)}|z_{ik}^{(t-1)}, \ Y^{(t-1)}\right) = q_{k_{12}}^{1-z_{ik}^{(t-1)}} \cdot q_{k_{22}}^{z_{ik}^{(t-1)}} . \qquad (10.21)$$

In the above equations, $\eta_i$ and $1 - \eta_i$ are node $i$ susceptibility of being influenced by its neighboring nodes and social independence parameters, respectively. As the susceptibility parameter $\eta_i$ increases and thus the independency parameter $1 - \eta_i$ decreases, node $i$ is being more susceptible to be influenced by its neighbors and vice versa. $\phi(i, t)$ is the set of node $i$ neighboring nodes at time step $t$ and is formally defined for directed networks as:

$$\phi(i, t) = \{j|j \in N, \ (j, \ i) \in E^{(t)}\}. \qquad (10.22)$$

Furthermore, $\omega_j \in \mathbb{R}^+$ indicates the influence weight of the neighbor node $j$ of the neighbor set of node $i$ that controls the influence on node $i$. In addition, $c_k \in \mathbb{R}^+$ and $b_k \in \mathbb{R}^+$ are scale coefficient and bias parameter of feature $k$, respectively.

### 10.4.3  Relationship between node features and link generation in dynamic social networks

This subsection describes the relationship between the node features and link generation process in social networks in the proposed model. To describe this part of the model, a stochastic matrix $P^{(t)}$ is defined that for any given pair of nodes $i$ and $j$ determines their interaction probability $p_{ij}^{(t)}$ based on their microscopic node features and the link matrix $Y^{(t)}$ as formally described as:

$$P^{(t)} = (p_{ij}^{(t)}) \in [0, 1]^{|N^{(t)}| \times |N^{(t)}|}. \qquad (10.23)$$

The proposed model is based on the idea of Multiplicative Attribute Graph Model [35,39,40]. As the authors in References 35, 39 and 40 describe, each feature $k$ is associated with a link affinity matrix, $\theta_k \in \mathbb{R}^{2 \times 2}$. Entries $\theta_{k_{11}}, \theta_{k_{12}}, \theta_{k_{21}}$ and $\theta_{k_{22}}$ of the matrix $\theta_k$ correspond to the tendency of linking between two nodes with the same features or with different features. These elements of the link affinity matrix are considered to be probability values.

Given feature vectors $z_{ik}^{(t)}$ and $z_{jk}^{(t)}$ of nodes $i$ and $j$ at time step $t$, the probability of generating link between these two node is defined by function $g(.)$ as:

$$p_{ij}^{(t)} = g\left(z_{i.}^{(t)}, \ z_{j.}^{(t)}\right). \qquad (10.24)$$

The quantities $z_{i.}^{(t)}$ and $z_{j.}^{(t)}$ can be applied as binary indicators of an entry $\theta_k = \left[z_{ik}^{(t)}, \ z_{jk}^{(t)}\right]$ of matrix $\theta_k$, for every feature $k$, at time step $t$. Therefore, the linking probability from node $i$ to node $j$ can be considered based on their features. Now, the stochastic matrix $P$ could be defined formally as:

$$p_{ij}^{(t)} = g\left(\sum_{k=1}^{\infty} \theta_k \left[z_{ik}^{(t)}, z_{jk}^{(t)}\right]\right),$$ (10.25)

where $g(x)$ is defined as:

$$g(x) = \frac{e^x}{1 + e^x}.$$ (10.26)

Nevertheless, considering an infinite number of features makes calculating the above summation impossible. As subtraction of $\theta_k[0, 0]$ from the other three entries of matrix $\theta_k$ and adding a bias parameter $\text{\ss}_t$ in the above equation, does not change the final summation, the above equation can be rewritten as:

$$p_{ij}^{(t)} = g\left(\text{\ss}_t + \sum_{k=1}^{\infty} \theta_k \left[z_{ik}^{(t)}, z_{jk}^{(t)}\right]\right),$$ (10.27)

$$\theta_k[0, 0] = 0.$$

Putting the three main processes explicitly described and modeled in Section 10.4 all together, Figure 10.2 graphically illustrates the proposed model for dynamic networks. As it is obvious from the figure that the model is an FHMM with hidden chain for variables $Z^{(t)}$ and $W^{(t)}$, and the observed

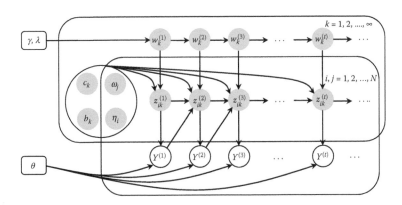

**FIGURE 10.2:** A graphical representation of the normal modeling component of the proposed approach. (Adapted from Y. Yasami & F. Safaei, 2017. *Computer Communications*, 100, 52–64. [44])

variables $Y^{(t)}$ that influence the hidden variables $Z^{(t)}$ and therefore cascade from observed to hidden variables.

---

## 10.5   Detecting Anomalous Entities

Now, after a detailed description of the infinite feature cascade model, it is possible to describe how to detect anomalous entities of dynamic networks. In the next subsections, the application of the proposed model in detecting anomalous links and nodes of dynamic networks is described.

### 10.5.1   Detecting anomalous links

For the purpose of detecting anomalous link in dynamic networks, a stochastic anomaly matrix $A^{(t)}$ for a special time step $t$ is defined based on the previously defined stochastic matrix $P^{(t)}$. The entry $(i, j)$ of $A^{(t)}$ can be described as:

$$A^{(t)} = \left( a_{ij}^{(t)} \right) \in [0, 1]^{|N^{(t)}| \times |N^{(t)}|},$$

$$a_{ij}^{(t)} = 1 - p_{ij}^{(t)}. \tag{10.28}$$

Each entry $p_{ij}^{(t)}$ of the stochastic matrix $P^{(t)}$ is defined as probability of normal interaction between node pair of $i$ and $j$. Thus, $A^{(t)}$ could be described as a probability of abnormal interaction between node pair of $i$ and $j$. A threshold rule with two threshold values $\tau_l$ and $\tau_h$ is applied to stochastic anomaly matrix $A^{(t)}$ that results in linking anomaly score matrix, $LAS^{(t)}$. The entry $(i, j)$ of this matrix is defined as:

$$LAS^{(t)} = (las_{ij}^{(t)}) \in \{0, 1\}^{|N^{(t)}| \times |N^{(t)}|},$$

$$las_{ij}^{(t)} = \begin{cases} \overline{Y}_{ij}^{(t)}, & \text{if } a_{ij}^{(t)} \leq \tau_l \\ 0, & \text{if } \tau_l < a_{ij}^{(t)} < \tau_h \\ Y_{ij}^{(t)}, & \text{if } a_{ij}^{(t)} \geq \tau_h. \end{cases} \tag{10.29}$$

Zeros in matrix $las_{ij}^{(t)}$ correspond to normal links, whereas ones correspond to abnormal links. Abnormal links are the network links that are not expected to exist but exist, as well as the network links that are expected to exist but do not exist.

## 10.5.2 Detecting anomalous nodes

To detect anomalous nodes, normalized node anomaly score $NAS^{(t)}$ is introduced, whose $i$th entry is defined formally as follows:

$$NAS^{(t)} = \left( nas_i^{(t)} \right) \in \{0, 1\}^{|N^{(t)}|},$$

$$nas_i^{(t)} = \frac{\sum_{j \in \phi'(i,t)} las_{ij}^{(t)}}{|\phi'(i, t)|}, \tag{10.30}$$

where $\phi'(i, t)$ for undirected networks can be defined as:

$$\phi'(i, t) = \{j | j \in N, Y_{ij}^{(t)} = 1\}. \tag{10.31}$$

Again, a threshold rule with a threshold value $\tau$ is applied to detect abnormal nodes as defined below:

$$i \sim \begin{cases} abnormal, & \text{if } NAS_i^{(t)} > \tau, \\ normal, & \text{if } NAS_i^{(t)} < \tau \end{cases}. \tag{10.32}$$

---

## 10.6 Model Inference and Experimental Results

This section provides the model inference, experimental setup, experimental results, and analysis on some real dynamic social networks.

### 10.6.1 Model inference

Some MCMC sampling strategies were applied to simulate parameters of the proposed model, given real social networks data. Each hidden feature chain $z_{ik}^{(1:t)}$ was sampled using the forward–backward recursion algorithm [41], given all the other model parameters and quantities. In deterministic forward pass defined by the algorithm, the chain is run down starting at the first time step. At each of the next time steps, the information collected from the data and parameters was stored in a dynamic programming cache, up to the last time step. In deterministic backward pass defined by the algorithms, each $z_{ik}^{(t)}$ is sampled using the information collected in dynamic programming cache, given all the network observations $Y^{(t)}$, in the reverse order, from the last time step to the first time. Sampling active features and link affinity matrix was performed using another MCMC strategy, Metropolis–Hastings algorithm [42].

### 10.6.2 Experimental setup

For anomaly detection experiments, a subset of the following two real directed social networks data sets was used [43]. First, the Google+ data set consists of "circles" from Google+. Google+ data were collected from users who had manually shared their circles using the "share circle" feature. The data set includes node features (profiles), circles, and ego networks. Second, the Twitter data set consists of "circles" (or "lists") from Twitter. Twitter data were crawled from public sources. The data set includes node features (profiles), circles, and ego networks.

For the purpose of link anomaly detection (LAD), the network observations $Y^{t_{tr}}$ throughout the entire training period time steps have been used as training data for each model. The aim was to detect abnormalities of network data $Y^{t_{tr}+1}$ during test period time step (time steps after the entire training period time steps) by each foresaid model. Therefore, about 1.5–2% of elements of $Y^{t_{te}}$ were complemented as abnormal links/non-links throughout the test period time step, that is, about 1.5–2% of links of the test data, $Y^{t_{te}}$, were converted to non-links and non-links of the test data, $Y^{t_{te}}$, were converted to links.

### 10.6.3 Experimental results and analysis

The evaluations were carried out upon the statistical performance measures of binary classification tests of normal/abnormal. These statistical measures, which have also been applied for anomaly detection in [44,45], are explained briefly as:

- *Sensitivity* (*True Positive Rate* or *TPR*): the probability that a test result is positive when there is anomaly.

- *Specificity* (*True Negative Rate* or *TNR*): the probability that a test result is negative when there is no anomaly.

- *Positive Likelihood Ratio* (*LR+*): the ratio between the probability of a positive test result, given the presence of the anomaly, and the probability of a positive test result, given the absence of the anomaly (*False Positive Rate, FPR*), that is:

$$LR+ = \frac{TPR}{FPR} = \frac{TPR}{1 - TNR}. \tag{10.33}$$

- *Negative Likelihood Ratio* (*LR−*): the ratio between the probability of a negative test result, given the presence of the anomaly, and the probability of a negative test result, given the absence of the anomaly, that is:

$$LR- = \frac{FPR}{TNR} = \frac{1 - TPR}{TNR}. \tag{10.34}$$

- *Positive Predictive Value* (*PV+*): the probability that the anomaly is present when the test result is positive.

- *Negative Predictive Value* (*PV−*): the probability that the anomaly is not present when the test result is negative.

The proposed anomaly detection approach was compared with four other approaches DMMG [34], LFP [12], DRIFT [10], and LFRM [2]. For this purpose, the models presented in [2,10,12,34] were applied for modeling the normal behavior of the network dynamics, during the entire training period time steps, while the anomaly detection approach presented in Section 10.5 was deployed during the test period time steps.

Diagrams presented in Figures 10.3 through 10.8 show the performance of the proposed approach in comparison with the other four approaches on both the Google+ and Twitter dynamic social networks data sets. Experiments were performed over four different test time steps and were repeated until the confidence interval of 95% was reached at each time step. As clearly shown in the figures, although the performance results were highly dependent on the data sets, but both the proposed link anomaly detection (LAD) and node anomaly detection (NAD) approaches had higher performance. On the other hand, the LFRM had tremendously lower performance characteristics in comparison with the other four approaches, in terms of all of the introduced performance measures.

Moreover, for both networks data sets and in all cases, the NAD approaches have lower performance compared to the LAD approaches. Furthermore, deep deliberation of the diagrams makes it clear that for both network data sets and in the two cases of LAD and NAD approaches, the proposed anomaly detection approach and LFP had increasing performance measures over the four test time steps, in terms of about all of the performance measures. Our argument is that as the number of training period time steps increases the model capability of anomaly detection in feature cascade increases.

As it can be seen in diagrams of Figure 10.3, the proposed LAD and NAD approaches had improved performance, in terms of *sensitivity* characteristics,

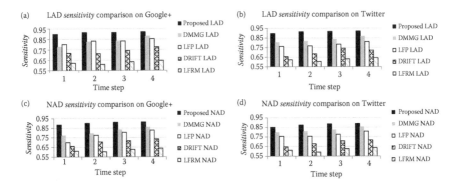

**FIGURE 10.3:** *Sensitivity* characteristics of the proposed anomaly detection approach. Results of (a) LAD approaches on Google+; (b) LAD approaches on Twitter; (c) NAD approaches on Google+; and (d) results of NAD approaches on Twitter.

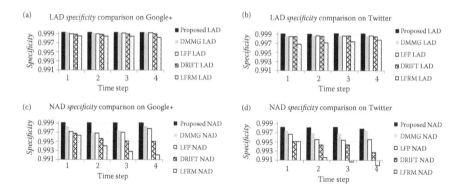

**FIGURE 10.4:** *Specificity* characteristics of the proposed anomaly detection approach. Results of (a) LAD approaches on Google+; (b) LAD approaches on Twitter; (c) NAD approaches on Google+; and (d) NAD approaches on Twitter.

outperforming the other approaches. DMMG and LFP performed better than DRIFT and LFRM. Experimental results also show low *sensitivity* measure for LFRM, too. The results obtained from both data sets are evidence on this statement. Moreover, LFP LAD approach was more sensitive than DMMG LAD approach in some time steps, on Google+ data set. While all the foresaid LAD and NAD approaches showed very high performance in terms of *specificity* and *PV−* measures, the comparative view of these performance measures is of great importance. As it is clear from each of diagrams illustrated in Figures 10.4 and 10.6 that the proposed approach outperformed others

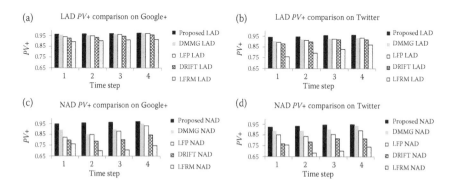

**FIGURE 10.5:** *Positive predictive value* characteristics of the proposed anomaly detection approach; Results of (a) LAD approaches on Google+; (b) LAD approaches on Twitter; (c) NAD approaches on Google+; and (d) NAD approaches on Twitter.

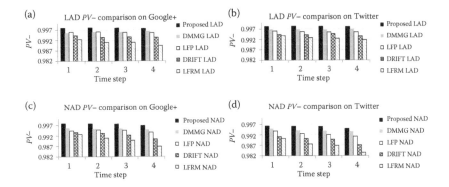

**FIGURE 10.6:** *Negative predictive value* characteristics of the proposed anomaly detection approach; Results of (a) LAD approaches on Google+; (b) LAD approaches on Twitter; (c) NAD approaches on Google+; and (d) NAD approaches on Twitter.

considering these measures, while DMMG, LFP, DRIFT, and LFRM LADs were placed in the next positions. Again, LFP LAD showed higher performance than DMMG LAD in some time steps for both data sets.

The proposed LAD and NAD approaches put in a better performance in terms of $PV+$, as indicated in diagrams illustrated in Figure 10.5. It was observed that LFRM performs extremely poorly compared to the other approaches, in term of this metric, too. Furthermore, this diagram indicates that LFP LAD outperformed DMMG LAD in most test time steps, on both data sets.

Diagrams depicted in Figures 10.7 and 10.8 reveal that the proposed LAD and NAD approaches tremendously outperformed the other four approaches on both the real dynamic social networks data sets, in terms of $LR+$ and $LR-$. Poor performance of LFRM LAD and NAD approaches is also obvious from these diagrams, too. Again, LFP LAD performed better than DMMG LAD in some time steps.

Table 10.2 shows the average estimated parameters for all of the mentioned approaches in all the cases for both dynamic networks data sets, over four test period time steps. As the table data indicates the proposed LAD and NAD approaches had higher performance characteristics than the other approaches. In terms of *sensitivity* performance measure, the proposed LAD approach had improved average performance of above 91% over both data sets, while the proposed NAD approach had the average performance characteristics of above 90% and 87% over Google+ and Twitter, respectively.

Average *specificity* and $PV-$ performance measures of the mentioned approaches also indicated better performance for the proposed LAD and NAD approaches. The proposed LAD approach had improved average

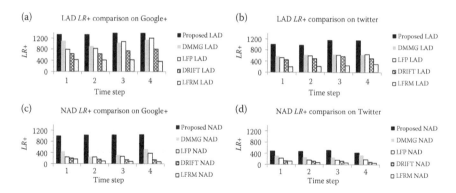

**FIGURE 10.7:** *Positive Likelihood Ratio* characteristics of the proposed anomaly detection approach; Results of (a) LAD approaches on Google+; (b) LAD approaches on Twitter; (c) NAD approaches on Google+; and (d) NAD approaches on Twitter.

$PV+$ of 97% on Google+ data set and above 95% on Twitter data set. This performance measure for the proposed NAD was above 96% and 94% over Google+ and Twitter, respectively. Improved average performance of the proposed approaches in terms of $LR+$ and $LR-$ is also notable. Lower average $PV-$ for the proposed LAD and NAD approaches indicated higher performance. The proposed LAD approach had reduced average $LR-$ of 8.7% on Google+ and 8.8% on Twitter. The improved average performance for the proposed NAD approach was 9.7% and 12.5% for Google+ and Twitter, respectively.

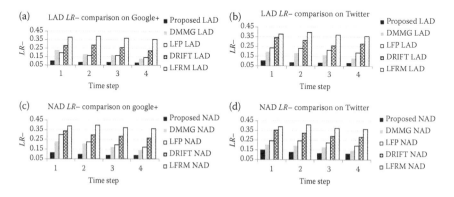

**FIGURE 10.8:** *Negative Likelihood Ratio* characteristics of the proposed anomaly detection approach; Results of (a) LAD approaches on Google+; (b) LAD approaches on Twitter; (c) NAD approaches on Google+; and (d) NAD approaches on Twitter.

**TABLE 10.2:** The average estimated parameters on real dynamic networks data sets: (a) the average estimated parameters for LAD and (b) the average estimated parameters for NAD

(a) 

| Parameters | LAD average estimated parameters on Google+ | | | | | LAD average estimated parameters on Twitter | | | | |
|---|---|---|---|---|---|---|---|---|---|---|
| | Proposed LAD | DMMG LAD | LFP LAD | DRIFT LAD | LFRM LAD | Proposed LAD | DMMG LAD | LFP LAD | DRIFT LAD | LFRM LAD |
| Sensitivity | 0.913 | 0.828 | 0.832 | 0.738 | 0.627 | 0.912 | 0.832 | 0.783 | 0.702 | 0.626 |
| Specificity | 0.99933 | 0.99920 | 0.99913 | 0.99895 | 0.99846 | 0.99914 | 0.99866 | 0.99869 | 0.99860 | 0.99730 |
| PV+ | 0.970 | 0.961 | 0.957 | 0.944 | 0.908 | 0.953 | 0.922 | 0.919 | 0.905 | 0.815 |
| PV− | 0.99789 | 0.99592 | 0.99595 | 0.99368 | 0.99095 | 0.99831 | 0.99679 | 0.99581 | 0.99426 | 0.99274 |
| LR+ | 1360.59 | 1039.68 | 980.68 | 708.29 | 409.08 | 1068.01 | 619.87 | 597.97 | 503.31 | 236.31 |
| LR− | 0.087 | 0.172 | 0.168 | 0.262 | 0.373 | 0.088 | 0.168 | 0.218 | 0.299 | 0.376 |

(b)

| Parameters | NAD average estimated parameters on Google+ | | | | | NAD average estimated parameters on Twitter | | | | |
|---|---|---|---|---|---|---|---|---|---|---|
| | Proposed NAD | DMMG NAD | LFP NAD | DRIFT NAD | LFRM NAD | Proposed NAD | DMMG NAD | LFP NAD | DRIFT NAD | LFRM NAD |
| Sensitivity | 0.903 | 0.815 | 0.777 | 0.705 | 0.621 | 0.875 | 0.822 | 0.775 | 0.688 | 0.618 |
| Specificity | 0.99912 | 0.99770 | 0.99724 | 0.99566 | 0.99384 | 0.99814 | 0.99719 | 0.99586 | 0.99419 | 0.99185 |
| PV+ | 0.963 | 0.898 | 0.873 | 0.809 | 0.730 | 0.940 | 0.905 | 0.862 | 0.799 | 0.723 |
| PV− | 0.99749 | 0.99530 | 0.99437 | 0.99236 | 0.99004 | 0.99584 | 0.99410 | 0.99249 | 0.98959 | 0.98708 |
| LR+ | 1028.59 | 383.48 | 288.74 | 166.44 | 109.85 | 473.98 | 295.79 | 189.95 | 120.33 | 81.80 |
| LR− | 0.097 | 0.186 | 0.224 | 0.296 | 0.382 | 0.125 | 0.178 | 0.226 | 0.313 | 0.386 |

## 10.7    Conclusions and Future Work

This chapter introduced the infinite feature cascade model and its application in detecting anomalous entities in dynamic social networks. The infinite feature cascade model initially models the normal dynamic behavior of dynamic networks over training period time steps. Modeling of each node features birth, longevity and death, dynamics of nodes features as well as its relationship with network links were taken into account by the proposed model. Subsequently, the deviation of network behavior was quantized using a novel approach in test period time steps. Experimental results on real dynamic social networks data sets showed that the presented approach for detecting anomalous links and anomalous nodes outperforms the other four related approach in terms of some statistical measures of performance specially applied in binary classification tests of normal/abnormal.

There are various interesting avenues for potential future work. First, this chapter merely studies the directed dynamic social networks. Still, the performance of the proposed model and anomaly detection approach needs to be evaluated on undirected dynamic social networks as well as other complex networks. Second, one of the major issues with the proposed approach is tuning of the threshold values $\tau_l$, $\tau_h$, and $\tau$. Therefore, application of other statistical and machine learning approaches needs to be studied for anomaly detection tasks.

## References

1. J. R. Lloyd, P. Orbanz, Z. Ghahramani, & D. M. Roy, 2012. Random function priors for exchangeable arrays with applications to graphs and relational data. *26th Annual Conference on Neural Information Processing Systems (NIPS)*, 3–8 December 2012, Harrahs and Harveys, Lake Tahoe, USA, 1007–1015.

2. K. T. Miller, T. L. Grifths, & M. I. Jordan, 2009. Nonparametric latent feature models for link prediction. *23th Annual Conference on Neural Information Processing Systems (NIPS)*, 7–9 December 2009, Vancouver, B.C., Canada, 1276–1284.

3. Y. Yasami & F. Safaei, 2018. A novel multilayer model for missing link prediction and future link forecasting in dynamic complex networks. *Physica A: Statistical Mechanics and its Applications*, 492, 2166–2197.

4. E. M. Jin, M. Girvan, & M. E. J. Newman, 2001. The structure of growing social networks. *Physical Review E*, 64, 046132.

5. A. L. Barabási, H. Jeong, Z. Néda, E. Ravasz, A. Schubert, & T. Vicsek, 2002. Evolution of the social network of scientific collaboration. *Physica A*, 311(3–4), 590–614.

6. D. Liben-Nowell & J. Kleinberg, 2003. The link prediction problem for social networks. *12th International Conference on Information and Knowledge Management (CIKM)*, New Orleans, LA, USA, 556–559.

7. Y. Yasami, 2017. A new knowledge-based link recommendation approach using a non-parametric multilayer model of dynamic complex networks. *Knowledge-Based Systems*, 143, 81–92.

8. E. M. Airoldi, D. M. Blei, S. E. Fienberg, & E. P. Xing, 2008. Mixed membership stochastic blockmodels. *Journal of Machine Learning Research*, 9, 1981–2014.

9. J. Yang & J. Leskovec, 2012. Community-affiliation graph model for overlapping community detection. *10th IEEE International Conference on Data Mining (ICDM)*, 10–13 December 2012, Brussels, Belgium, 1170–1175.

10. J. Foulds, A. U. Asuncion, C. DuBois, C. T. Butts, & P. Smyth, 2011. A dynamic relational infinite feature model for longitudinal social networks. *14th International Conference on Artificial Intelligence and Statistics (AISTATS)*, 11–13 April 2011, Fort Lauderdale, USA, 287–295.

11. F. Guo, S. Hanneke, W. Fu, & E. P. Xing, 2007. Recovering temporally rewiring networks: a model-based approach. *24th International Conference on Machine Learning (ICML)*, 20–24 June 2007, Corvallis, Oregon, USA, 321–328.

12. C. Heaukulani & Z. Ghahramani, 2013. Dynamic probabilistic models for latent feature propagation in social networks. *30th International Conference on Machine Learning (ICML)*, 16–21 June 2013, Atlanta, GA, USA, 275–283.

13. P. Sarkar & A. W. Moore, 2005. Dynamic social network analysis using latent space models. *19th Annual Conference on Neural Information Processing Systems (NIPS)*, 5–8 December 2005, Vancouver, B.C., Canada, 1145–1152.

14. S. Kairam, D. Wang, & J. Leskovec, 2012. The life and death of online groups: Predicting group growth and longevity. *5th ACM International Conference on Web Search and Data Mining (WSDM)*, 8–12 February 2012, Seattle, WA, USA, 673–682.

15. N. Shrivastava, A. Majumder, & R. Rastogi, 2008. Mining (social) network graphs to detect random link attacks. *24th IEEE International Conference on Data Engineering (ICDE)*, April 7–12, 2008, Cancún, México, 486–495.

16. W. Fu, L. Song, & E. P. Xing, 2009. Dynamic mixed membership blockmodel for evolving networks. *26th International Conference on Machine Learning (ICML)*, 14–18 June 2009, Montreal, Quebec, Canada, 329–336.

17. Q. Ho, L. Song, & E. P. Xing, 2011. Evolving cluster mixed-membership blockmodel for time-varying networks. *14th International Conference on Artificial Intelligence and Statistics (AISTATS)*, 11–13 April 2011, Fort Lauderdale, USA, 342–350.

18. K. Ishiguro, T. Iwata, N. Ueda, & J. Tenenbaum, 2010. Dynamic infinite relational model for time-varying relational data analysis. *24th Annual Conference on Neural Information Processing Systems (NIPS)*, 6–11 December 2010, Hyatt Regency, Vancouver Canada, 919–927.

19. T. Griffths & Z. Ghahramani, 2006. Infinite latent feature models and the Indian buffet process. *Advances in Neural Information Processing Systems*, 18, 475–482.

20. N. A. Heard, D. J. Weston, K. Platanioti, & D. J. Hand, 2010. Bayesian anomaly detection methods for social networks. *The Annals of Applied Statistics Institute of Mathematical Statistics*, 4(2), 645–662.

21. R. Hassanzadeh, R. Nayak, & D. Stebila, 2012. Analyzing the effectiveness of graph metrics for anomaly detection in online social networks. *Lecture Notes in Computer Science: Web Information Systems Engineering*, 7651, 624–630.

22. Z. Chen, W. Hendrix, & N. F. Samatova, 2012. Community-based anomaly detection in evolutionary networks. *Journal of Intelligent Information Systems*, 39(1), 59–85.

23. S. Hanneke, W. Fu, & E. P. Xing, 2010. Discrete temporal models of social networks. *Electronic Journal of Statistics*, 4, 585–605.

24. T. A. B. Snijders, G. G. van de Bunt, & C. E. G. Steglich, 2010. Introduction to stochastic actor-based models for network dynamics. *Social Networks*, 32(1), 44–60.

25. M. S. Handcock, G. Robins, T. Snijders, J. Moody, & J. Besag, 2003. Assessing degeneracy in statistical models of social networks. *Journal of the American Statistical Association*, 76, 33–50.

26. T. Snijders, 2006. Statistical methods for network dynamics. *2006 Proceedings of the XLIII Scientific Meeting*, Italian Statistical Society. Luchini, S. R. (ed.). Padova: CLEUP, p. 15p.

27. T. Snijders, 2001. The statistical evaluation of social network dynamics. *Sociological Methodology*, 31(1), 361–395.

28. D. Crandall, D. Cosley, D. Huttenlocher, J. Kleinberg, & S. Suri, 2008. Feedback effects between similarity and social influence. *14th ACM SIGKDD International Conference on Knowledge Discovery and Data Mining (SIGKDD)*, 24–27 August 2008, Las Vegas, Nevada, USA, 160–168.

29. P. D. Hoff, A. E. Raftery, & M. S. Handcock, 2002. Latent space approaches to social network analysis. *Journal of the American Statistical Association*, 97(460), 1090–1098.

30. E. P. Xing, W. Fu, & L. Song, 2010. A state-space mixed-membership block model for dynamic network tomography. *Annals of Applied Statistics*, 4(2), 535–566.

31. A. H. Westveld & P. D. Hoff, 2011. A mixed effects model for longitudinal relational and network data, with applications to international trade and conflict. *Annals of Applied Statistics*, 5(2A), 843–872.

32. J. Van Gael, Y. W. Teh, & Z. Ghahramani, 2008. The infinite factorial hidden Markov model. *22nd Annual Conference on Neural Information Processing Systems (NIPS)*, 8–11 December 2008, Vancouver, B.C., Canada, 1697–1704.

33. P. Sarkar, S. M. Siddiqi, and G. J. Gordon, 2007. A latent space approach to dynamic embedding of co-occurrence data. *10th International Conference on Artificial Intelligence and Statistics (AISTATS)*, 21–24 March 2007, San Juan, Puerto Rico, 420–427.

34. M. Kim & J. Leskovec, 2013. Nonparametric multi-group membership model for dynamic networks. *27th Annual Conference on Neural Information Processing Systems (NIPS)*, 5–10 December 2013, Harrahs and Harveys, Lake Tahoe, USA, 1385–1393.

35. M. Kim & J. Leskovec, 2012. Latent multi-group membership graph model. *29th International Conference on Machine Learning (ICML)*, June 26–July 1, 2012, Edinburgh, Scotland, UK.

36. M. Mørup, M. N. Schmidt, & L. K. Hansen, 2011. Infinite multiple membership relational modeling for complex networks. *21st IEEE International Workshop on Machine Learning for Signal Processing (MLSP)*.

37. S. J. Gershman, P. I. Frazier, & D. M. Blei, 2015. Distance dependent infinite latent feature models. *IEEE Transaction on Pattern Analysis and Machine Intelligence*, 37(2), 334–345.

38. E. Meeds, Z. Ghahramani, R. Neal, & S. Roweis, 2006. Modeling dyadic data with binary latent factors. *20th Annual Conference on Neural Information Processing Systems (NIPS)*, 4–7 December 2006, Vancouver, B.C., Canada, 977–984.

39. M. Kim & J. Leskovec, 2011. Modeling social networks with node attributes using the multiplicative attribute graph model. *27th Conference on Uncertainty in Artificial Intelligence (UAI)*, 14–17 July 2011, Barcelona, Spain, 400–409.

40. M. Kim & J. Leskovec, 2012. Multiplicative attribute graph model of real-world networks. *Internet Mathematics*, 8(1–2), 113–160.

41. S. L. Scott, 2002. Bayesian methods for hidden markov models. *Journal of the American Statistical Association*, 97(457), 337–351.

42. S. Chiband & E. Greenberg, 1995. Understanding the Metropolis-Hastings algorithm. *The American Statisticians*, 49(4), 327–335.

43. http://snap.stanford.edu/.

44. Y. Yasami & F. Safaei, 2017. A statistical infinite feature cascade-based approach to anomaly detection for dynamic social networks. *Computer Communications*, 100, 52–64.

45. Y. Yasami & S. Pourmozaffari, 2009. A novel unsupervised classification approach for network anomaly detection by k-means clustering and ID3 decision tree learning methods. *Journal of Supercomputing*, 53(1), 231–245.

# Chapter 11

# Finding Justice through Network Analysis

Radboud Winkels

## CONTENTS

## 11.1   Introduction

In the last decade, more and more legal documents have become publicly available online in many countries throughout the world. It often concerns both legislation and case law and sometimes other documents like legal commentaries. Professionals may already get lost in the multitude of information, let alone ordinary citizens and small-to-medium-sized enterprises. Besides the quantity of information, there are some other problems:

- The documents are not always nicely structured, with sections or paragraphs that can be referred to. They may vary from bitmaps to well-structured XML documents.

- Links within the document and references to other (parts of) documents are not always explicit and machine readable; especially across data collections (from different sources) and across jurisdictions.

265

- Documents and parts of documents do not always have unique and permanent identifiers, or ways of identifying them, which also makes it hard to link them in a systematic and sustainable way (see previous point).

In most (Western) countries, commercial publishers traditionally provide legal information and support for access. Legal experts write commentaries, editors provide links between different (types of) sources and point subscribers to interesting new developments and case law. Now that a large amount of sources of law become electronically available online for free, the question is whether new ways for supporting access can be developed. One stream of research may be directed towards *crowdsourcing*, where users of legal information share their collections of material, the links they see between different sources, their commentaries and so on. Another stream of research is directed at (semi-) automated linking and clustering of sources of law, analysis of the network of law to find authoritative sources or predict the change of opinion of higher courts, and so on.

In the OpenLaws.eu project we explored both approaches [1]. We developed a platform that enables users to find legal information more easily, organize it the way they want and share it with others. The primary focus is on legal professionals, but ultimately the platform should be useful for everybody. The OpenLaws Internet platform is based on open data, open innovation, and open source software.

In this article, I will focus on what we call *automated enrichment* of legal data: using techniques like machine learning, network analysis, and natural language processing to automatically create new metadata for sources of law. I will describe several experiments in the field of Dutch immigration law where we do this in an attempt to create a *Legal Recommender System* [2].

The structure of this paper is as follows: First, I will describe how we create a web of law if it is not available in machine-readable form, or extend it when that is necessary. Next, I will explain how we recommend sources of law to users given a current document they are investigating based on network analysis. The focus will be on legislation and case law. In the next section, we will add other document features, particularly "text similarity", to suggest interesting new documents. I will also present small formative evaluations of all approaches and end with conclusions and further research.

## 11.2   Creating a Web of Law

As stated above, sources of law form a network of (parts of) documents; they cite and reference each other. We distinguish three types of documents with different characteristics:

*Legislation:* Legal rules of a generally binding character, typically have explicit processes for declaring rules in force and for modifying existing rules. Typically has a complex reference structure. The bibliographic distinction between "works" and "expressions" is of critical importance, because the text changes over time [3]. "Expressions" are the different versions of a "work", for example, "article x of law y in force at date *t*" and the same article at date "*t* + 1" where the text differs. Figure 11.1 presents an example of a legislative network.

*Case Law:* Legal decisions on individual cases that have a formative influence on future decisions in similar cases. This is most obviously the case for the judiciary, but other arbitration bodies may in practice be relevant in everyday life (for instance, arbitration courts for consumer complaints). These documents are typically never changed over their lifetime and do not have to be declared in force or retracted. Legal decisions may be selectively published, however, depending on their formative influence on future decisions. Cases cite other cases and—depending on the legal system—legislation.

*Commentaries:* Documents that function as an expert commentary on legal rules and decisions. These documents typically make references to rules or decisions, but otherwise, share little in terms of structure or metadata with sources of law. Products of legal publishers fall under this category.

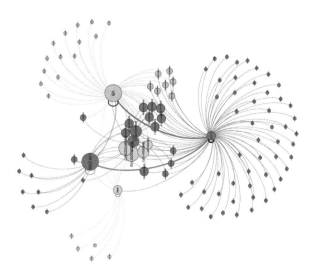

**FIGURE 11.1:** The network of Dutch Immigration Law. Shadings denote different laws (e.g., light gray is "Vreemdelingenbesluit" (VB)); the size of nodes the relative importance of the law based on a number of ingoing and outgoing references. The references are depicted as lines between nodes.

Several researchers have applied network analysis to legal documents, but never for recommending new documents and only to one type of data at a time: for establishing the authority of case law as described in [4] and [5], or analysing the structure of legislation as described in [6] and [7]. van Opijnen [8] uses links to legislation in Dutch case law when deciding upon the relevance of a particular case, but not to suggest other relevant sources of law. In our work, we use network analysis of both legislation and case law to recommend potentially interesting new sources of law.

### 11.2.1   Legal portals

Most of the sources of law available online are stand-alone web services or databases, containing one type of documents, not linked to other sources. For instance, the Dutch portal for case law contains a (small) part of all judicial decisions in the Netherlands. Case citations in these decisions are sometimes explicitly linked, references to legislation are not. From earlier research, we know that professional users of legal documents would like to see and have easy access to related ones from other collections. For example, when we evaluated a prototype system that recommends other relevant articles and laws to users of the official Dutch legislative portal, experts told us that they would like to see relevant case law and parliamentary information as well [9]. In this paper, we present a step in that direction. The new version of our portal presents relevant case law, given a legislative article in focus for a user, and adapts the ranking of relevant other articles based on the related case law. The idea is that judges in explaining and justifying their verdicts—applying the law in practice—indicate that the sources they cite are somehow related.

Another problem of existing portals and data bases is that not even all internal links are explicitly represented. This is especially true for so-called relative links like "the previous article," or "the second sentence of article x" and incomplete ones ("that law" or "article y" without the law that it is part of).

If these links are not given, we can try to find them automatically. For inter- and intra-legislation links we have shown this can be done very effectively for the Dutch case [10] and others for other jurisdictions (e.g., [11] for Italy; [12] for Japan). For inter-case law links, it is a bit more difficult, but it works for the Dutch case [5,8]. That leaves finding citations in case law to legislation and possibly finding links to other sources of law like commentaries to both case law and legislation. Here we will discuss finding references to legislation in Dutch case law and how we can use these to improve access to Dutch sources of law.

### 11.2.2   The Dutch case law portal

The Dutch portal *rechtspraak.nl* contains a small but growing part of all judicial decisions in the Netherlands. At the time of this research, case citations in these decisions were sometimes explicitly marked in metadata (e.g., the first

instance case); references to legislation only the main one(s) in recent cases.[1]
The texts were available in an XML format, basically divided into paragraphs
using tags, with a few metadata elements. The most relevant metadata for
our purpose are:

- The date of the decision ("Datum uitspraak").

- The field(s) of law ("Rechtsgebieden").

- The court ("Instantie").

The court decisions did not contain inline, explicit, machine-readable links
to cited legislation or other cases. So even when the metadata contained such
references, we do not know in which paragraph the case or article was cited.
We resorted to parsing techniques to make these citations explicit and count
them. We chose to work with a subset of all case law to start with; those cases
that were tagged as belonging to "immigration law" and that contained the
actual text of the verdict. That gave us 13,311 documents to work with.

For locating references to legislation in case law we used regular expressions, as we have done in the past, together with a list of names and abbreviations of Dutch laws [10]. This list also contains the official identifier of the
law (the BWB number), which can be used for resolving the reference later
on. We consider high precision to be more important than high recall. Users
will forgive us if we miss a reference, but be annoyed by false ones. We evaluated this procedure by checking 25 randomly selected documents by hand.
These documents contained 163 references to legislation of which 141 were
correctly identified (recall of 87%). There was one false positive (precision
of 99%). The references we missed were mostly those to the "*Vreemdelingen-
circulaire*" ("Aliens act", a lower law that has a different structure than regular ones) and treaties with very long names like "*Europees Verdrag tot
bescherming van de rechten van de mens en de fundamentele vrijheden*"
("Convention for the Protection of Human Rights and Fundamental Free-
doms"). When we tried to capture this with regular expressions, they would
match too easily, often matching entire sentences where they should have
matched only the law. We declared these conventions outside the scope of
this first experiment.

Resolving the references was a bit trickier, since sometimes they used anaph-
ora, for example, referring to "that law." In such cases, the citation was resolved
by using the previous law identifier if it existed, that is, we assume that the complete law was introduced just before in the text and resolved correctly. We used
the same process for resolving ambiguous title abbreviations; for example,
"WAV" is an abbreviation of "*Wet Arbeid Vreemdelingen*" (law on labour for

---

[1]Recently, the portal was upgraded and now also provides better structured linked data.

immigrants), "*Wet Ammoniak en Veehouderij*" (law on ammonia and livestock) and "*Wet Ambulancevervoer*" (law on ambulance transport). Most of the time the full title is used before the abbreviation is used.

Another issue is determining the exact version of the law the case refers to. Typically, a judge will refer to the version that is in force at the moment of the decision, but it may also be the version that was in force at the time of the relevant facts, or even sometimes an earlier version of the relevant law and so on. We cannot decide which version is the correct one without interpreting the content of the case. Therefore, we decided to resolve the reference to the "work" level of the source of law, that is, no particular version (see Figure 11.2). The resulting references are added to the XML of the case law document. The final network of the 13,311 case documents has 85,639 links to legislation (on average 6.5 references per case); the links connect the ECLI identifier of the case with the BWB identifier of the (part of) law it refers to.

We evaluated the resolving process by checking 250 random ones by hand. Of these, 234 should have been resolved since the other 16 were outside of the scope of this experiment. One hundred and ninety-eight were resolved correctly (a recall of 85%). We had 10 "false" positives, that is, references that were declared out of scope, so a precision of 95%. The results were good enough to continue.

Since case decisions may refer to the same source of law, for example, an article, more than once, we count the number of references and compute the weight of the link between the case and the article as $W = 1/n$, where n is the number of occurrences of a certain reference and $W$ is the weight of the edge. The lower the weight, the stronger the impact on the network is.

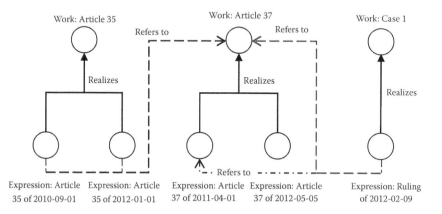

**FIGURE 11.2:** Bibliographic levels of documents and referencing. Article 35 has two expressions, both refer to the work level of article 37. A decision Case 1 (the only expression of the work) refers to article 37, given the date probably to the first expression, but we represent it as referring to the work level (light gray arrow).

## 11.3   Recommending Sources of Law

When a user selects an article in our prototype recommender portal, related case law and legislation are retrieved (see Figure 11.3):

1. The system checks whether the article appears in the case law network. If so, it creates a so-called ego graph, a local network containing all the nodes and edges within a certain weighted distance from the current node [13]. We start searching with a weighted distance of 0.4 and gradually increase it up to 2.0 until we have a sufficiently large, but still manageable network.

2. To find relevant legislation, the system also checks whether the current node is in the legislative network. If so, it again creates an ego graph, this time for an unweighted network. To control the size of the graph, we use only references coming from the selected version (expression) of the current node.

3. If we have two local networks, we want to combine them in order to (better) predict the importance of legislative nodes. To do this, we need to assign weights to the legislative graph. We chose the value 0.1 as it allows the legislative network to influence the result but not overrule the case law references.

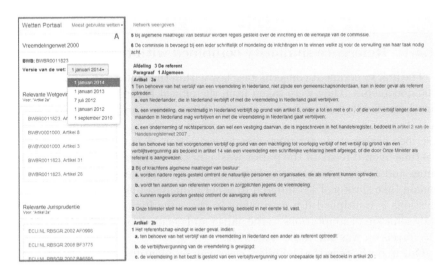

**FIGURE 11.3:** The user has article 2a of the Immigration law in focus. The current version is January 2014 (see pull-down menu at left). Relevant other articles are presented in window A (dark border) on the left and below that relevant case law.

4. Finally, we use betweenness centrality on the combined network to determine the most relevant articles for the current focus. Betweenness centrality is a measure of centrality in a graph based on shortest paths. For every pair of nodes in a graph, there exists at least a shortest path between the nodes such that either the number of edges that the path passes through (for unweighted graphs) or—as in this case—the sum of the weights of the edges (for weighted graphs) is minimized.

5. The results are shown to the user in frame A of Figure 11.3.

## 11.4     Other Document Features

In the previous experiment, we only exploited the web of references between documents. Now we will explore whether we can improve suggestions for relevant documents by including other features, notably similarity measures based on the comparison of the actual text of documents. As stated in the Introduction, we focus on case law in this study.

Relevancy for case law is hard to define; it is subjective, depends on the task or problem of the user searching for case law and also on the type of user. For a student or novice, well-known landmark cases might be very relevant, while for a legal expert these are probably not. He or she will be more interested in less known or very recent new cases. In previous research, we have concentrated on legal expert users and we still are, but we will also have a look at the preferences of novices.

### 11.4.1     Reference similarity combined with text similarity

The first experiment concerns case law within the Dutch tax domain, about 6000 documents. They were taken from the official Dutch portal based on the metadata "field(s) of law" (see above). After some preprocessing in which XML tags and strange characters were removed, we used the same parser as described earlier to detect references to legislation. A small test revealed that this time overall recall of the parser was only 55%. The main reason is missing names and abbreviations of tax laws and the use of different terms in the tax domain (like "protocol") compared to other domains of law. The effect of these omissions is multiplied if a case refers to the same item more than once. We made some small adaptations to the parser and decided to continue the research since precision was still more than 95%.

Our hypothesis is that "similar" cases can be identified by similarity in the legislation they cite and by similarity of the words used in the judgments. As a baseline, we use the similarity in words used (text similarity): Bag-of-words combined with normalized TF/IDF weighting and cosine similarity. TF/IDF corrects for the frequency of terms and normalization for the length of documents; otherwise, documents with highly frequent words and long documents would be overrated by the cosine similarity measure. For any document in our set (the focus document) we can now select the $n$-most similar other documents from that set. The left part of Table 11.1 gives an example of the 10 most similar documents found this way for a verdict of the court of Amsterdam of 2010 (BO1378) on the value of a house. The most similar case is one of the court of Alkmaar of 2008 (BC6103), also on the value of a house.

The same algorithms are used to calculate the reference structure similarity between two documents. The right part of Table 11.1 gives the 10 most similar documents based on that method. Here number 1 is a verdict of the court of The Hague of 2008 (BD1495) on pollution tax for commercial property.

The final step combines the bag-of-words similarity score and the bag-of-references similarity score by taking the average of the two scores. Now we can determine the $n$ most similar documents for any *focus* document. We chose to only take into account the focus documents that have at least four outgoing references for now, this because of the relatively low recall of the parser. If a focus document has only one outgoing reference, the bag-of-references similarity scores will be either 0.0 or 1.0; this may be acceptable in a later stage with a superior reference parser, but at this stage, these extreme scores are too uncertain.

### 11.4.2 Formative evaluation

A small group of experts was asked to evaluate the system. They were asked to first read a focus document, randomly selected from a prepared database and shown on an evaluation website that was created for this purpose. Subsequently, they were asked to read six recommended documents and rank them on relevancy to the focus document. The most relevant one is ranked first (1), the least relevant last (6). The six documents were three with the highest similarity scores for the baseline implementation (bag-of-words only) and three documents with

**TABLE 11.1:** Ten most similar documents to RBAMS-2010-BO1378 according to bag-of-words (left) or bag-of-references (right)

| | ECLI-NL-RBAMS-2010-BO1378 | | | |
|---|---|---|---|---|
| | **Bag-of-words** | | **Bag-of-references** | |
| 1 | RBALK-2008-BC6103 | 0.72 | RBSGR-2008-BD1495 | 0.89 |
| 2 | RBDOR-2010-BM0117 | 0.69 | RBUTR-2010-BU4490 | 0.73 |
| 3 | RBALK-2011-BQ0469 | 0.64 | RBOVE-2014-951 | 0.69 |
| 4 | RBARN-2006-AY9465 | 0.64 | RBALK-2008-BD7537 | 0.69 |
| 5 | RBDOR-2010-BO5257 | 0.62 | RBALK-2007-BB9105 | 0.69 |
| 6 | RBAMS-2011-BV6758 | 0.61 | RBAMS-2011-BQ424 | 0.65 |
| 7 | RBDOR-2010-BM2339 | 0.61 | RBALK-2008-BC4175 | 0.64 |
| 8 | GHAMS-2013-CA2684 | 0.61 | RBALK-2012-BX0044 | 0.59 |
| 9 | RBAMS-2011-BR6478 | 0.60 | RBALK-2008-BD5937 | 0.58 |
| 10 | RBHAA-2006-AZ2187 | 0.59 | GHAMS-2001-AD8208 | 0.57 |

the highest similarity scores for the bag-of-words combined with the bag-of-references. They were also asked to give an overall score for the relevance of a suggestion on a scale from 1to 10 with 1 representing "not relevant" and 10 "very relevant." This was done to assess the overall quality of suggestions. Even very bad suggestions may be ranked after all. Our intuition was that it is easier to rank suggestions for relevance than to assess the overall relevance of a suggestion.

This method of evaluation is rather subjective if a small amount of test subjects is used, but it can serve to determine if the research is heading in the right direction and find possible bugs and errors [14].

Four experts evaluated 18 cases; 15 unique ones and 3 the same for all experts. Results are presented in Table 11.2. It is clear that adding the bag-of-references worsens performance. This may partly be due to the fact that only 55% of references were found.

It may also be that similarity in references reflects something else than similarity in the text. To further investigate this aspect, we examined to what extend the bag-of-words approach suggests the same documents as the bag-of-references approach. We analyzed the top 10 recommendations for 1000 focus documents for both approaches. We ignored recommendations with a similarity score of zero. Of the remaining 9528 recommendations, 1135 were recommended by both algorithms (about 12%).

Again, the amount of overlap may turn out to be higher with a better performing parser. Another explanation is that similarity in references reflects a more abstract commonality than our evaluators could spot or found useful.

### 11.4.3    Network analysis combined with topic modeling

A more advanced approach of comparing the similarity of texts than the bag-of-words approach discussed above is that of topic modeling. A topic model represents a document, a court judgment in this case, as a mixture of topics. A topic is a set of words or phrases. Perhaps the most common topic model currently in use is Latent Dirichlet Allocation (LDA) [15]. One downside of this approach is that it treats documents again as bags of words, that is, ignores word order and therefore word phrases like "European Union" or "our Minister." Several extensions of LDA have been proposed, one of which is Turbo Topics [16] which starts like LDA, but subsequently significant words that are preceding or succeeding topic-words are searched in the texts and added to the topic

**TABLE 11.2:**    Results expert evaluation

|                                        | Average rank | Average score |
| -------------------------------------- | ------------ | ------------- |
| Bag-of-words only (baseline)           | 3.06         | 5.58          |
| Bag-of-words with bag-of-references    | 3.94         | 4.46          |

model. In this experiment, we use the open source implementation of Turbo Topics in MALLET.

To be able to compare the results with suggestions purely based on network analysis, we decided to use the same domain as in the earlier work described above, namely immigration law. After some preprocessing and selection of those cases with actual content, we worked with a set of nearly 13,500 cases.

After all cases were represented as mixtures of topics, we selected the $n$ best suggestions for each case by calculating the similarity between the topic mixtures. We calculate the sum of squared errors between a specified case and all other cases and convert this to a similarity value ranging from 0% to 100% overlap. Table 11.3 gives an example for three suggestions for case ECLI:NL: RBSGR:2009:BH7787 (a case of 2009 of the court of The Hague about a refugee from Turkey who was a courier for PKK) with similarity measures. It obviously has a similarity measure of 100% with itself. Next comes a case from the same court of 2004 with a similarity in topics of more than 99% (also about a member of the PKK) and so on.

### 11.4.4 Formative evaluation

For this project, the evaluators consisted of three novices, two legal experts with experience in the immigration law, and two legal experts without experience in this field. For the results, the four legal experts were pooled together, since there was no difference in their evaluations. The evaluators were given five randomly selected cases for which they ranked three suggestions from best to worst, and for which they stated whether any of the three suggestions are good enough for a recommender system. One of the three suggestions was most similar according to the method using a topic model, another was the most relevant based on the references to legislation according to and one of the suggestions was based on a combination of the two methods. For the combination of the two methods, a list of the top 200 suggestions according to the topic model was obtained (for each randomly selected case), and also a list of the top 10 suggestions according to the references to legislation. The first suggestion from the top 10 list that appeared in the other list of the top 200 suggestions, was chosen as the suggestion based on both methods. The evaluators were unaware of which suggestion was obtained by which method.

**TABLE 11.3:** Example output of three suggestions for case ECLI:NL: RBSGR:2009:BH7787

| Case | Similarity measure |
| --- | --- |
| ECLI:NL:RBSGR:2009:BH7787 | 100.00 |
| ECLI:NL:RBSGR:2004:AQ5970 | 99.28 |
| ECLI:NL:RBDHA:2015:4915 | 99.01 |
| ECLI:NL:RVS:2004:AQ5615 | 98.67 |

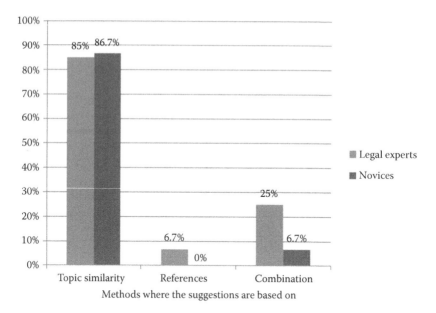

**FIGURE 11.4:** Percentage of evaluators that thought a suggestion was good enough for a recommendation.

Both legal experts and novices showed a significant preference for the suggestions based on topic similarity. The legal experts wanted to see 85% of the suggestions based on topic similarity in a recommender system, for the novices this was 87% (see Figure 11.4). The legal experts ranked the suggestions based on topic similarity as best suggestion 80% of the time, while the novices always ranked the suggestion based on topic similarity as the best suggestion. This indicates that suggestions based on topic similarity can give useful suggestions within Dutch case law.

## 11.5    Conclusions

We described results of research that is part of the OpenLaws.eu project. Our ultimate aim is to deliver a platform for using, sharing, and enriching big open legal data. Besides offering users the opportunity to collect, organize, and annotate legal data, we also want to offer automatic suggestions based on analysis of existing data. In the future that may also be based on user-generated data. Users of OpenLaws can already annotate sources of law, highlight sections or phrases, group sources together in folders and link a source to other sources. In the future, these data can be analyzed as well to improve suggestions for other users.

For now, the recommendations are based on the analysis of official sources of law. We have presented first results for available data in the Netherlands. Based on the analysis of the network of both legislation and case law, we offer users suggestions for interesting material given their current focus document.

We have shown that it works quite well to automatically find and resolve references to legislation in Dutch case law. The parser was perhaps a bit over-fitted for the immigration domain since it performed less well in the tax domain. It can easily be improved, but we will have to check whether this has repercussions for the immigration domain and how it performs in other legal fields. We did not exploit the network of case law itself this time. Earlier [5] we showed that this can be used to estimate the authority of cases, so including this may improve the relevance of suggested case law.

The network of references can be used to provide users of the legislative portal with relevant judicial decisions given their current focus and moreover, suggest additional relevant legislative sources. Another step is adding legal commentaries and doctrine to the network and possibly parliamentary data.

When compared with suggestions of case law based on similarity of the actual text of the judgments, whether seen as just "bags of words" or as a mixture of "topics", users seem to prefer those over the suggestions based on network analysis or similarity in reference structures. We still have the intuition that similarity in reference structure indicates some common feature of cases, but perhaps it is too abstract for users. In our first experiments, we treated the references as an unordered set; perhaps we should retain the order and try again.

In the future, we will use some additional features as:

- The relative frequency of the reference in a court decision (what van Opijnen calls "multiplicity" [8]).

- The hierarchical position of the law cited, for example, whether the referred law is a European directive or treaty, or a governmental decree.

- Document structure level of the reference. A lower document structure level (e.g., article or clause instead of a chapter) suggests a more specific reference, which could indicate a different role.

- The date of a case, preferring more recent cases for expert users. We can assume that experts are well aware of older (landmark) cases, and will be more interested in (very) recent ones.

---

# Acknowledgments

Part of this research is co-funded by the Civil Justice Programme of the European Union in the OpenLaws.eu project under grant JUST/2013/JCIV/

AG/4562. I would like to thank my students: Erwin van den Berg, Bart Vrede-bregt and Wolf Vos who performed the experiments.

# References

1. Wass, C., Dini, P., Eiser, T., Heistracher, Th., Lampoltshammer, Th., Marcon, G., Sageder, C., Tsiavos, P. and Winkels, R. 2013. OpenLaws.eu. In: Proceedings of the 16th International Legal Informatics Symposium IRIS 2013, Salzburg, Austria.

2. Winkels, R., Boer, A., Vredebregt, B. and Van Someren, A. 2014. Towards a legal recommender system. In Legal Knowledge and Information Systems: JURIX 2014, the Twenty-seventh Annual Conference, pp. 169–178. IOS Press. Best paper award.

3. Saur, K. G. 1998. Functional requirements for bibliographic records. *UBCIM Publications – IFLA Section on Cataloguing*, 19:136.

4. Fowler, J.H. & Jeon, S. 2008. The authority of Supreme Court precedent, *Social Networks* 30: 16–30, Elsevier.

5. Winkels, R.G.F., de Ruyter, J. & Kroese, H. 2011. Determining authority of Dutch case law. In K. Atkinson (ed). *JURIX 2011*, 103–112. IOS Press, Amsterdam.

6. Liiv, I., Vedeshin, A. and Täks, E. 2007. Visualization and structure analysis of legislative acts: A case study on the law of obligations. *ICAIL 2007*, pp. 189–190, ACM, New York.

7. Mazzega, P., Bourcier, D. and Boulet, R. 2009. The network of French legal codes. In *ICAIL 2009*, pp. 236–237.

8. van Opijnen, M. 2014. Open in het web. Hoe de toegankelijkheid van rechterlijke uitspraken kan worden verbeterd. *PhD Thesis in Dutch*, University of Amsterdam.

9. Winkels, R.G.F., Boer, A. and Plantevin, I. 2013. Creating context networks in Dutch legislation. In K. Ashley (ed). *JURIX 2013*, 155–164. IOS Press, Amsterdam.

10. de Maat, E. Winkels, R., and van Engers, T. 2006. Automated detection of reference structures in law. In T. van Engers (ed), *JURIX 2006*, 41–50. IOS Press, Amsterdam.

11. Palmirani, M., Brighi, R. and Massini, M. 2003. Automated extraction of normative references in legal texts. In: 9th International Conference on AI and Law, 105–106, ACM, New York.

12. Tran, O.T., Le Nguyen, M. and Shimazu, A. 2013. Reference resolution in legal texts. In: Proceedings of the Fourteenth International Conference on Artificial Intelligence and Law, 101–110, ACM, New York.

13. Newman, M. 2010. *Networks: An Introduction*. Oxford, England: Oxford University Press.

14. Shani, G. and Gunawardana, A. 2011. Evaluating recommendation systems. In: *Recommender Systems Handbook*, pp. 257–297. Springer, Berlin.

15. Blei, D.M., Ng, A.Y., Jordan, M.I. 2003. Latent Dirichlet allocation. *Journal of Machine Learning Research* 3, 2003, 993–1022.

16. Blei, D. and Lafferty, J.D. 2009. Visualizing topics with multi-word expressions. *arXiv*, 0907, 2009.

# Index